Experimental Phycology 1

The proceedings of the previous two Algae Conferences were published as:

Compartments in Algal Cells
Edited by W. Wiessner, D. G. Robinson, and R. C. Starr
(1983) ISBN 3-540-13318-6

Algal Development
Edited by W. Wiessner, D. G. Robinson, and R. C. Starr
(1986) ISBN 3-540-17716-7

Experimental Phycology

Cell Walls and Surfaces, Reproduction, Photosynthesis

Edited by
W. Wiessner, D.G. Robinson,
and R.C. Starr

With 91 Figures

Springer-Verlag
Berlin Heidelberg New York
London Paris Tokyo
Hong Kong Barcelona

Professor Dr. Wolfgang Wiessner
Professor David G. Robinson
Pflanzenphysiologisches Institut der Universität
Untere Karspüle 2
3400 Göttingen, FRG

Professor Richard C. Starr, Ph. D.
The University of Texas at Austin
Austin, TX 78712, USA

ISBN 3-540-52496-7 Springer-Verlag Berlin Heidelberg New York
ISBN 0-387-52496-7 Springer-Verlag New York Berlin Heidelberg

Offsetprinting: Druckhaus Beltz, Hemsbach/Bergstr.
Bookbinding: Schäffer, Grünstadt
2131/3145-543210 – Printed on acid-free paper

Preface

Phycology in Göttingen has enjoyed a long and illustrious history. It began over a 100 years ago with Johannes Reinke and has continued in this century in the hands of Richard Harder, Ernst Georg Pringsheim and André Pirson. We are proud that algae have now become "institutionalized" in Göttingen in the form of a thriving culture collection and through the existence of a Department in the Plant Physiology Institute devoted to experimental phycology.

Phycology has been additionally embedded in the academic landscape of Göttingen through a number of successful international meetings. Since 1950 these symposia, which have taken place at 3 year intervals, have been organized under the auspices of the Akademie der Wissenschaften in Göttingen with generous financial support of the Volkswagen Foundation. With the help of Springer Verlag we have been able to publish the proceedings of two of these meetings. This "tradition" is continued with the present publication, which includes most of the papers delivered at the 1989 Göttingen symposium.

In comparison to previous meetings which were devoted to a central theme, for example in 1983 "Compartments in Algal Cells and their Interaction" and in 1986 "Algal Development", the September 1989 meeting was devoted to several, different areas of algal research. This led us to an even better exchange of ideas and information among the participants than had occurred in the past. Because of this, and since we are planning to hold future meetings in this manner, we have decided to "institutionalize" the title of the proceedings. This book therefore carries the name "Experimental Phycology, Vol. 1".

We hope that it will be just as successful as its predecessors. Thanks to the efforts of the editorial staff of Springer Verlag this volume now appears, once again, within only a few months of the meeting. Its contents are therefore highly topical.

In closing we would like to express our thanks to the external members of our advisory committee (Professors Melkonian, Schnepf and Sitte), and to all those in Göttingen who helped in running this symposium.

Göttingen, April 1990

W.Wiessner
D.G.Robinson
R.C.Starr

Contents

Contributors

Addresses are to be found at the beginning of each article.

The emergence of eukaryotic algae within the protists: A molecular phylogeny based on ribosomal RNA sequencing

R. Perasso, A. Baroin, A. Adoutte
Laboratoire de Biologie Cellulaire 4
(URA 1134 du CNRS), Bâtiment 444
Université Paris-Sud
91405 Orsay-Cedex
France

A few words about molecular phylogenies.

The comparative analysis of homologous sequences of informational macromolecules (*i.e.* proteins or directly DNA) is providing a remarkably powerful tool to establish phylogenetic relationships among living organisms. Knowledge of the sequence of emergence of biological groups in turn illuminates a large number of key biological issues since, as stated long ago by Th. Dobzhansky, "Nothing in biology makes sense except in the light of evolution". There are several advantages to sequence data as compared to morphological, physiological or biochemical ones for establishing phylogenetic relationships :

- the sequencing of a single set of homologous genes or gene products in different species can yield numerous characters simultaneously. In fact each nucleotide (or amino acid) is potentially susceptible of corresponding to a character difference. Thus tens or even hundreds of characters can be screened through the comparative analysis of one large gene ;

- these characters evolve, to a first rough approximation, independently of each other *i.e.* substitution at one site is not necessarily related to substitution at another site. This is more difficult to evaluate for morphological characters whose correlated divergence may simply mean that they are under the control of the same genetic modification ;

- molecular characters are easier to quantify and, eventually, to weigh than morphological ones ;

- finally, and possibly most importantly, sequencing data can be obtained, at least for some widespread genes, even when morphology is exceedingly reduced or deformed ! (*i.e.* prokaryotes, poorly differentiated eukaryotes, parasites, and even DNA containing organelles).

This does not mean that molecular approaches should replace all others for establishing phylogenies since they also have a few pitfalls as will be seen below. What is exceedingly rewarding is the confrontation of the molecular data with the other types of information ; in fact it is mainly through the congruence of independent approaches that convincing phylogenetic conclusions are reached.

Two key issues in molecular phylogenetics are 1) the choice of the appropriate molecule to sequence and 2) the use of the best techniques to "extract" phylogenetically significant information from sequence comparison and to construct valid evolutionary trees. The work of Woese and his co-workers (see Woese, 1987 and Gray, 1988 for recent reviews) has clearly established the remarkable properties of the large ribosomal RNAs (*i.e.* 16-18S and 23-28S rRNA, often named SSU rRNA and LSU rRNA for small and large subunit ribosomal RNAs, respectively) as universal tools for constructing molecular phylogenies, specially when very distant species are involved since the molecules display long nucleotide stretches of an exceptional degree of conservation. This allows the alignment of homologous portions with no or very little ambiguity among species as distant as *E. coli* and *H. sapiens* and thus the construction of very broad phylogenies. Since these highly conserved stretches alternate with less conserved ones, comparison of more closely related groups can also be carried out. In addition ribosomal RNAs are universally distributed (including in mitochondria and chloroplasts) ; they carry the same function, have a large information content, and are abundant (which makes them easy to sequence directly by a DNA-primed reverse transcriptase method : Qu *et al.*, 1983 ; Lane *et al.*, 1986).

As for the methods of evolutionary trees reconstruction, they belong to two broad classes : the so-called "distance" methods and the parsimony ones. Their basis, domains of application, merits and limitations including testing by simulation (Sourdis and Krimbas, 1987 ; Sourdis and Nei, 1988 ; Saitou and Imanishi, 1989) have been discussed in detail in several recent methodological and review papers (Felsenstein, 1982, 1983, 1988 ; Lake, 1987 ; Li *et al.*, 1987 ; Olsen, 1987, 1988 ; Penny *et al.*, 1987 ; Cedergren *et al*, 1988 ; Nanney et al. 1989) .

Basically, distance methods start by calculating a value reflecting the amount of difference occurring between each of the pairwise combinations of sequences analyzed. This can be either a direct measure of the number of substitutions or can include various types of corrections to take into account the possibility of multiple substitutions at each site (see for example Kimura, 1981). One should also, in principle, take into account the fact that different nucleotide positions, along the length of the molecule under study, can display differences in their mutation rate. This, however, has been seldom considered so far (but see Olsen, 1987). A crude or corrected distance matrix is thus obtained. Then, most programs attempt to fit the topology of a network as well as the lengths of the various branches of the network to the distance values of the matrix, using one of several possible optimization criteria. A network is thus obtained which can be polarized to become a phylogenetic tree by choosing the position of a root. The first description of such a method is in a classic paper by Fitch and Margoliash (1967) and several of the distance-matrix programs use their original algorithm and optimization criterion. It is important to note that this class of distance methods does not assume the identity of mutation (fixation) rates among the various lineages under analysis, that is they do not postulate the existence of a universal molecular clock.

Distance methods thus reduce the information that was contained in aligned sequences to a single value between each pair of sequences. Parsimony methods, in contrast, retain the site to site information and attempt to construct the network linking all the sequences which requires the smallest number of mutational steps along its branches. By comparison with the approaches based on morphological characters, distance methods are related (although not identical) to phenetics while parsimony is in resonance with cladistics. Obviously, both in distance and parsimony methods, even powerful computers cannot examine all possible trees when the number of species rises above 12-15 and therefore choices have to be made as to how exhaustive the search is and/or to find astute alternatives to exhaustive searches.

Several algorithms and the corresponding computer programs corresponding to these approaches have been made available by their authors. A highly comprehensive and well documented package is that of Felsenstein ("Phylip"). A very efficient parsimony program is "PAUP" of D. Swofford. Both have been used in the present work. A recent addition to the distance methods is the "Neighbor-joining" strategy of Saitou and Nei (1987) which displays good efficiency as compared to the "Fitch" type of approach. It has also been used here.

As for the evaluation of the reliability of a topology, the mathematics are only starting. At least four types of parameters have to be considered, two related to the real historical biological situation one is trying to reconstruct and two to the quality of the sequence data and of the sampling : first, if one is dealing with a set of species that have diverged from each other during a relatively short period of time, one runs the risk of reaching a topology which resembles a bush : all the species branches will emerge from a set of poorly resolved common short stems. In such cases, the difficulty in deriving the true branching order is not due to a methodological problem but to a situation which is intrinsically difficult to solve. Unfortunately, such situations are often the most interesting ones biologically ! The debate over the branching order of man, chimpanzee and gorilla is probably an example of such a situation (see Holmquist *et al.*, 1988 a and b for a thorough discussion of all these problems). Second, one may be dealing with species that have experienced highly different mutation rates. This will introduce topological error in most constructions (Felsenstein, 1978) unless there is excellent additivity of values in distance matrix data or if special, recently developed methods, such as "evolutionary parsimony" (Lake, 1987) are used in parsimony type constructions. Third, one may have choosen sequences that undergo an inappropriate rate of mutation for the sample of species under analysis such, for example, that due to an excessive substitution rate, noise and homoplasies (identical sites due to "convergent" mutations and not to true homology) will be high. Again, there are risks of reaching wrong topologies in such cases. Finally, the molecule or portion of molecule choosen may be too short to obtain statistically significant information at least over the range of species choosen. Each of these difficulties can be evaluated within a given data set but not yet totally quantified. Thus, an empirical (but very useful) criterion of reliability, that of "robustness", is often used : confidence in a topology can be gained when it is stable in the face of modifications of the tree construction method, length of sequence considered, sample of species choosen within each group, etc... Two other strong empirical tests are the congruence of a molecular phylogeny with one well established on anatomical and/or paleontological bases. Vertebrates are obviously good test organisms to check that a given macromolecule and/or a given tree construction method behave properly ! The other test is the congruence of phylogenies based on different molecules.

All our work so far has been based on the analysis of the 450 nucleotides located at the 5' end of the LSU rRNA, obtained by the direct sequencing method (Qu *et al.*, 1983). Although shorter than complete SSU rRNA, this portion of the LSU rRNA has a number of very favorable features : it consists of two highly conserved domains bracketting a domain of greater variability. With respect to the reliability criteria just discussed, the conserved domains have proven remarkably efficient. They allow inter-phylum and even inter-kingdom comparisons that are both internally consistent and largely superposable with those obtained by other methods or other molecules (in particular with complete SSU and LSU rRNA). In addition the data base for this region is now quite substantial, comprising over 130 species. The length of this portion of the LSU rRNA may, however, be somewhat insufficient to conclusively resolve the topology of closely located branching points and caution must therefore be exercised when interpreting such situations.

Protists and algae : who they are and why study their phylogeny ?

One of our first objectives has been to reconstruct, the phylogeny of unicellular eukaryotes ("protists") ; taken in the broad sense, this is an extremely diversified assemblage of a large number of both photosynthesic and non-photosynthetic organisms, possibly amounting to 45 different phyla and at least to 18 "super-assemblages" (Corliss, 1984). To cite some, the protists include the ciliates, the rhizopods ("amoeba", "Acrasia", etc...) the dinoflagellates, the actinopods (Heliozoa, Acantharia, ...), the chlorophytes, the rhodophytes, the chromophytes, the cryptomonads, the euglenozoa, the polymastigotes (*i.e.* "flagellates" such as *Trichomonas*), the microsporidia, the sporozoa and even some organisms traditionally classified with the fungi !

As can be clearly seen from the above list, the modern understanding of the term "protist" has erased the traditional frontiers between organisms that belonged to the "botanical" world and those that belonged to the "zoological" one. In particular, eukaryotic algae which comprise a large number of unicellular representatives, have been grouped with the protists. Incidently, this creates some nomenclature problems. We have used a semi-conservative procedure, naming the taxa either according to the zoological or the botanical code so as to retain familiar names at lower taxonomic levels. In most cases we have used Corliss' (1984) terminology which is basically the traditional ones, except for a few simplifications at higher taxonomical levels.

Interest in protists is largely due to the idea that unicellular eukaryotes have played a key role in evolution as intermediates between prokaryotes and multicellular plants and animals. Assuming that present day protists are descendants of such "hinge" organisms, reconstructing their phylogeny should allow the clarification of two major evolutionary transitions : the prokaryote (or "progenote" (Woese, 1987)) to eukaryote one and the unicellular eukaryote to metozoa and metaphytes one. For example, one could try to identify the extant protist group whose point of emergence on a dendrogram is the closest to the metazoa or to the metaphytes root and thus obtain an idea of what the unicellular ancestor(s) of the multicellular kingdoms looked like. Conversely, one could try to identify the earliest emerging protist group and thus gain some insight into the early processes of eukaryogenesis. This is strikingly exemplified by the recent papers of Vossbrinck *et al.* (1987) and Sogin *et al.* (1989). One could also try to clarify the links that exist between these diverse unicellular groups, find possible grouping among them, determine their respective order of emergence and thus gain light into the processes that have governed evolution within the unicellular framework. These are the aims that we have set to our work and it is in this context that we have been led to analyze the algae.

The specific problems of algae phylogeny.

Eukaryotic algae are traditionally separated into three broad divisions : the rhodophytes (Chlorophyll a + phycobilins), the chromophytes (chlorophylls a + c) and the chlorophytes (chlorophylls a + b). Thus, following Christensen (1964), Whatley (1981) and most textbooks (see for example South and Whittick, 1987), taxonomical subdivision of the algae is mainly based on the photosynthetic and accessory pigments *i.e.* a chloroplastic trait. This does fit quite well with other properties of the plastids (reviewed by Gibbs, 1970 and Whatley, 1981). Algae, however, have in fact three distinct genetic compartments : the nuclear, mitochondrial and chloroplastic ones. For each of these compartments the question arises as to whether it is of a unique origin (monophyletic or polyphyletic ?) and what its relations to the equivalent compartment in other groups is (*i.e.* what is the closest nuclear relative of algae, for example ?).

The symbiotic origin of chloroplasts *i.e.* the settling of a photosynthetic bacterium in the cytoplasm of a eukaryotic host, can now be considered as established (Margulis, 1981 ;

Gray and Doolittle, 1982 ; Hunt *et al.*, 1985 ; Giovannoni *et al.*, 1988). All present day photosynthetic eukaryotes need not, however derive from a single symbiotic event. First, several distinct prokaryote-eukaryote associations may have occurred. Thus, chlorophytes and rhodophytes can be seen as resulting from either a single acquisition event or from two independent events. In the former case, the introduction of a cyanobacterium into a eukaryote giving rise to rhodophytes is followed by loss of phycobilins and development of chlorophyll b biosynthesis in a sub-line to yield chlorophytes. In the latter case, rhodophytes and chlorophytes originate from the introduction of distinct symbionts (a cyanobacterium and a prochlorophyte, for example) into distinct hosts (Lewin, 1983 ; Burger-Wiersma *et al.*, 1986). The second complication corresponds to the potential settling of a photosynthetic eukaryote in another eukaryote's cytoplasm. Such a scenario is now favored for chromophytes and cryptophytes (Dodge, 1979 ; Gibbs, 1981 ; Whatley, 1981 ; reviewed in Cavalier-Smith, 1986). The evidence resides principally in the presence of a unique and conspicuous system (the chloroplast endoplasmic reticulum, CER) made up of two concentric membranes surrounding the chloroplast in addition to the chloroplast envelope proper. The outermost of these two membranes bears ribosomes on the surface facing the cell's cytoplasm and in most representatives of the chromophytes as well as in the cryptophytes, is in topological continuity with the outer membrane of the nuclear envelope. Cavalier-Smith (1986) has argued convincingly that this outer CER membrane is derived from fusion of the phagosomal membrane with the endoplasmic reticulum of the host while the inner one would correspond to the symbiont's plasma membrane. Cryptophytes contain, in addition, within the space comprised between the CER and the chloroplast envelope, a mass of DNA surrounded by a typical nuclear envelope (the nucleomorph) which is thought to correspond to the residual nucleus of the endocytized photosynthetic eukaryote. Here again, the question can be raised as to whether chromophytes and cryptophytes derive from the same host and the same initial eukaryotic symbiont.

The broad questions concerning algal phylogeny can therefore be summarized as follows:

1 - Was symbiosis an early or a late event in unicellular eukaryotes history ?

2 - Are chloroplasts of mono- or polyphyletic origin ?

3 - Did a single or multiple eukaryotic species serve as host(s) for the symbiotic event(s) and what were their relation (s) with other protozoa (for instance "zooflagellates") ?

The mono or polyphyletic origin of chloroplasts can be addressed through the sole comparative analysis of chloroplastic sequences with respect to prokaryotic ones. By contrast, the two other questions require an analysis of nuclear encoded sequences, so as to evaluate the phylogenetic diversity of present day photosynthetic eukaryotes and examine whether or not they form a unique and coherent group relative to non-photosynthetic ones. For example, should the sequences of representatives of the chlorophytes and the rhodophytes lead to dendrograms in which the two groups are quite independent, this would reinforce the hypothesis of multiple symbiotic events. The same reasoning can be used to ascertain the common or independent origin of the eukaryotic symbiont of chromophytes and cryptophytes.

We have approached these questions by incorporating 10 partial sequences of nuclear LSU rRNA of photosynthetic protists into our previous database which consisted of non-photosynthetic organisms (Baroin *et al.*, 1988). The newly sequenced species correspond to one rhodophyte (a Bangiophyceae), five chromophytes (two Chrysophyceae, two Haptophyceae and a Rhaphydophyceae), two chlorophytes (a Chlorophyceae and a Prasinophyceae) and two representatives of the cryptophytes. A brief analysis of the results has been recently published (Perasso et al., 1989). Here, using an expanded set of tree building methods, we discuss these results in more detail and confront them to recent ideas and data from the literature.

Global topology of the dendrogram : late emergence of algae.

The ten new sequences aligned with those of representatives from the protists and the other kingdoms are shown in figure 1.These sequences were analyzed both by distance and by parsimony programs. Three distinct distance matrix methods were used ; the first (Fitch and Margoliash, 1967) is the classical least-square one ("Fitch" program of Felsenstein's Phylip computer package). The second is a recent procedure developed by Hénaut and Delorme (1988) which also starts from the distance data but treats them according to a very different algorithm, derived from dynamic clustering procedures which allows the identification of "strong clusters"in a multidimensionnal space. The third is the neighbor joining method of Saitou and Nei (1987) as programmed by D. Higgins (University of Dublin). For the parsimony approach, we have used both Swofford's "PAUP" program and Felsenstein's "DNA parsimony" one (discussed in Fink, 1986). For the four types of treatments, we have been restrictive in our choice of the domains considered as homologous (*i.e.* allowing unambigous alignment). These are overligned in figure 1. In so doing, we use the most highly conserved stretches of the RNA molecule.

Overall, there is a good agreement between the trees obtained with the five methods. However one frequent discrepancy in the topologies is noted between the distance trees and the parsimony ones, concerning the rhodophytes and cryptophytes and this will be discussed below. For the discussion of the global topology, we shall use the distance trees for three main reasons. First, these trees are quite "robust" *i.e.* their topology is not perturbed by addition of new species, slight modification of the length of the sequence stretches taken into account, etc... Second, the available length of sequence, the rates of substitution recorded and the broadness of the phylogenetic range under analysis, all suggest that distances may better exploit the data than parsimony (see Sourdis and Nei, 1988 for justifications). We

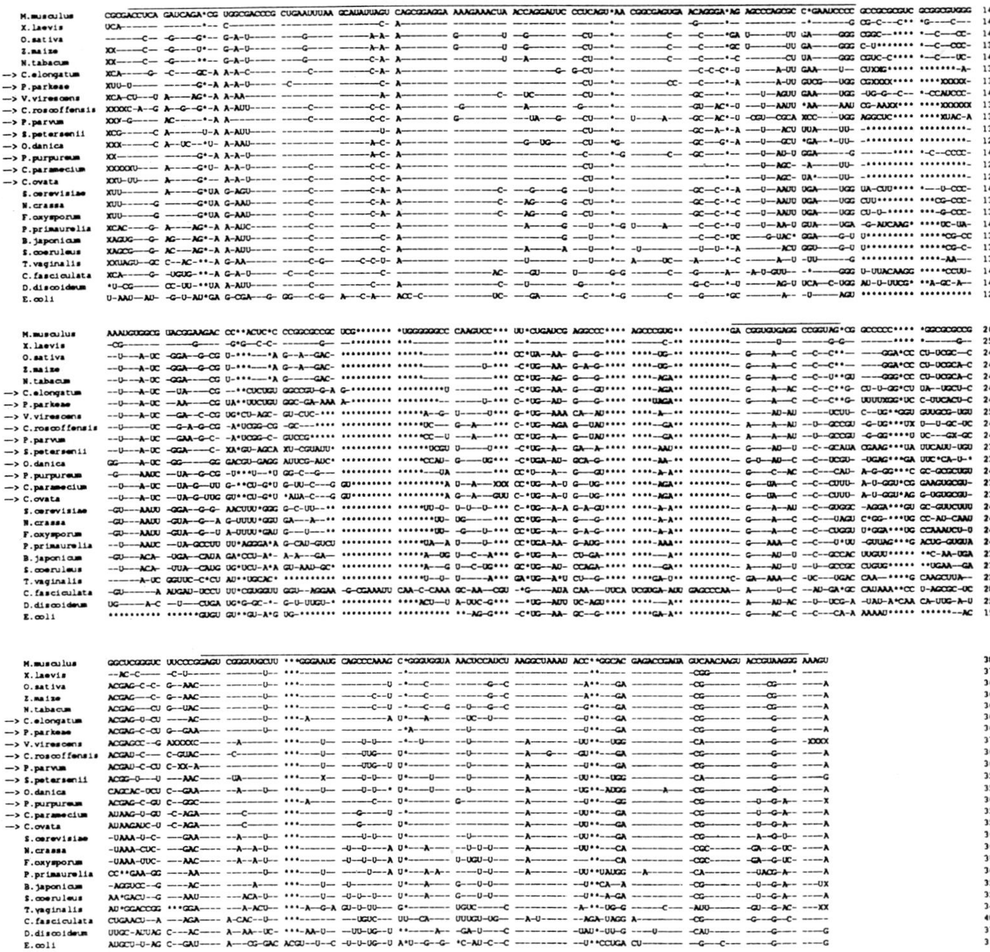

Figure 1
Sequence alignments for the 5'-terminal region of nuclear encoded large subunit rRNA. The photosynthetic eukaryotic species that we have sequenced are indicated by arrows ; the other sequences serve as references. Only the nucleotides that differ from those of the mouse are shown (identities are denoted by hyphens and deletions are denoted by stars). The nucleotide positions that could not be identified are denoted by an "X". References for the sequence data of species not described in this paper can be found in Baroin *et al* (1988).

have observed, for example, that parsimony trees display several unstable features with the number of species and type of sequences used in the present work. Even when the range of species was slightly reduced, instabilities persisted, most likely as a result of the very broad evolutionary array of phyla analyzed. Finally, focusing on distance methods allows easier comparison both with our previous data (Baroin *et al.*, 1988) and with those of other groups working on related organisms (Sogin *et al.*, 1986 ; Hori and Osawa, 1987 ; Vossbrinck *et al.*, 1987).

Figures 2, 3 and 4 depict the inferred evolutionary relationships of algae with the four eukaryotic kingdoms : metazoa (*Mus musculus* and *Xenopus* laevis), metaphytes (*Oryza sativa, **Zea mays*** and *Nicotiana tabacum*), fungi (*Saccharomyces cerevisiae* and *Fusarium oxysporum*) and the broad assemblage of protists. The protists representatives include three ciliates (*Stentor coeruleus*, *Blepharisma japonicum* and *Paramecium primaurelia*), a dinoflagellate (*Prorocentrum micans*), a slime mold (*Dictyostelium discoideum*), a euglenozoan flagellate (*Crithidia fasciculata*) and a parabasalid flagellate (*Trichomonas vaginalis*). The prokaryote *E. coli* serves as an outgroup. A number of very robust clusters are consistently observed with the three distance methods and whether the distance matrix is based on crude or corrected values. These are the group of chromophytes, that of chlorophytes and higher plants, that of ciliates, and that of metazoa.These clusters were also consistently obtained with parsimony methods. *Crithidia, Trichomonas,* and *Dictyostelium*

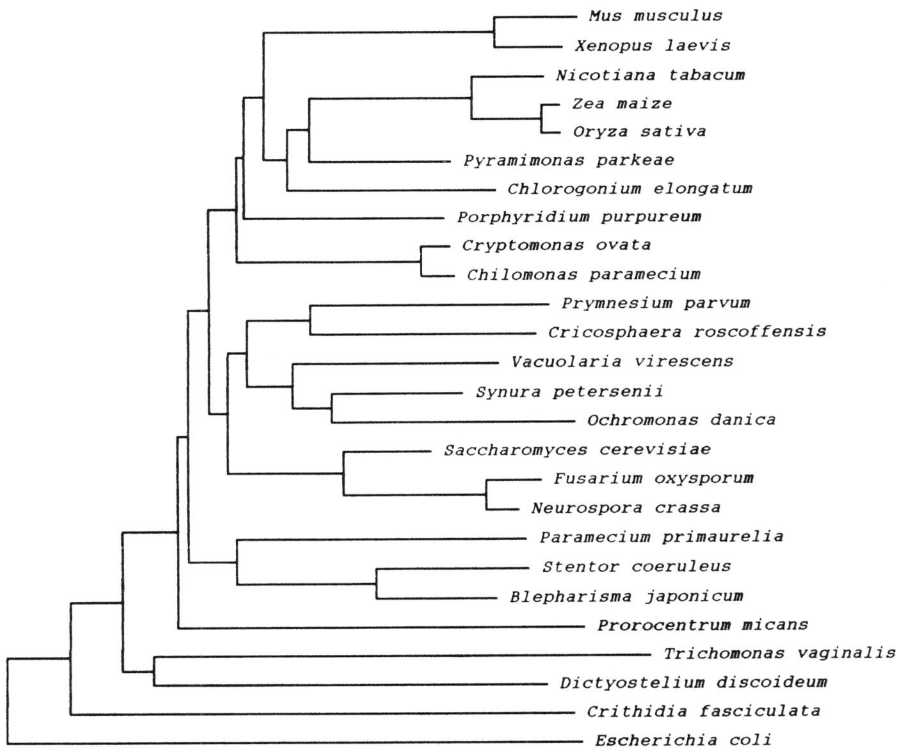

Figure 2

Phylogenetic tree constructed by Fitch and Margoliash's method using the "Fitch" program of Felsenstein Phylip package with the G (global rearrangement) option. Pairwise comparisons of the rRNA sequences have been carried out over the conservative regions overlined in Figure 1 and the uncorrected number of substitutions was used. In this tree, the distance between two species is proportional to the sum of the projections on the x axis of the branch lengths, while distances on the y axis are arbitrary. 2488 trees were examined and the average percent of standard deviation was 5.59.

always emerge clearly before all these groups. The three fungi are also strongly linked to the chromophytes. Among the less stable positions are that of the single dinoflagellate (*Prorocentrum*) which oscillates between ciliates and chromophytes, and those of rhodophytes and cryptophytes (see below).

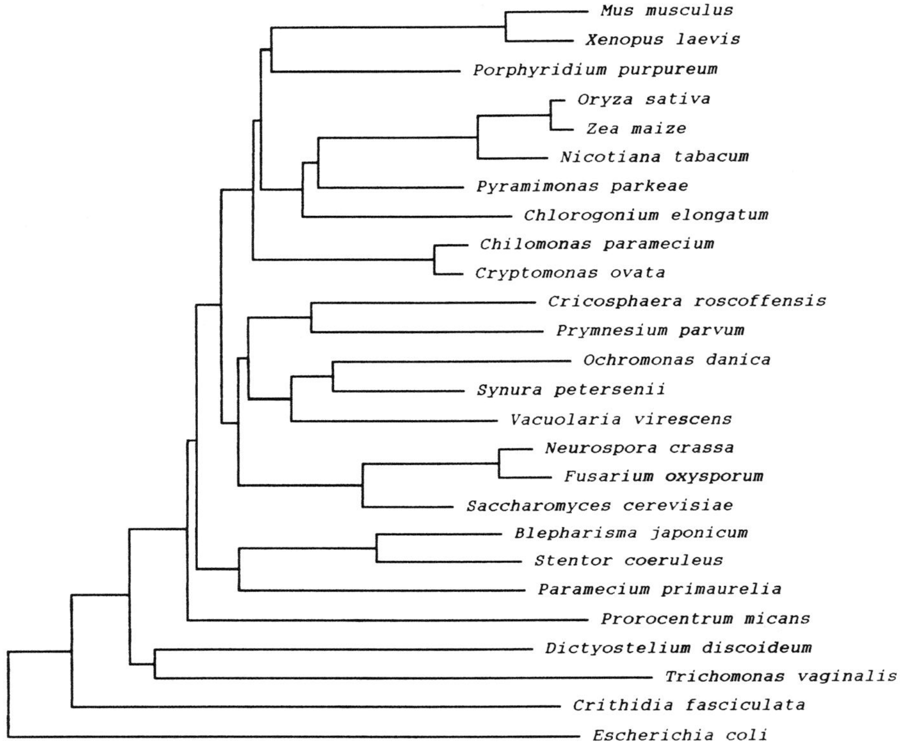

Figure 3
Phylogenetic tree constructed using the Neighbor Joining method (Saitou and Nei, 1988) as programmed by Higgins. Note the great similarity with the topology in Fig. 2 except for the position of the red algae.

The major point to be made from these global dendrograms is that all the algal groups emerge later than several representatives of the unicellular eukaryotes. Both Sogin's group (1986) and our's (Baroin *et al.*, 1988) had previously reported that a set of non-photosynthetic flagellated protists emerge very early from the eukaryotic tree. It now appears that the three major classes of photosynthetic protists as well as the cryptophytes have emerged at a later stage, during a period of intense diversification that is relatively close to the metazoa-metaphyte radiation and which comprises the fungi and other groups of protists such as ciliates and dinoflagellates. This is in marked contrast with the widespread view of an ancient origin of algae, particularly the rhodophytes. Red algae are often considered as a very primitive group mainly because of the cyanobacterial-like composition of their chloroplasts, their unstacked thylakoids, their lack of flagella and centrioles and their apparent structural simplicity. In all our dendrograms, however, rhodophytes are clearly posterior to many other protist groups. Indeed, they are even posterior to other algal groups in most of the distance trees and in some parsimony treatments. In fact, no unambiguous paleontological data in favor of a very ancient origin of rhodophytes have been reported so far. On the whole, the view that we favor in the light of the presently available evidence is that of a relatively long history of non-photosynthetic eukaryotic protists (see also Vossbrinck *et al.*, 1987 and Sogin *et al.*, 1989) followed by an explosive period closely predating the radiation of plants and animals. This may correspond in chronological terms to the late Precambrian.

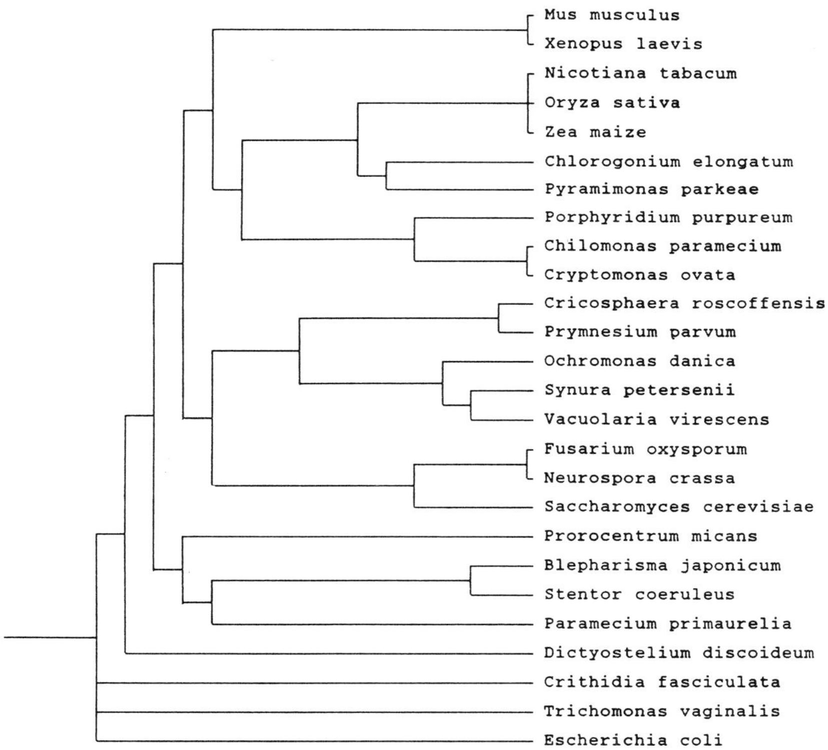

Figure 4

Dendrogram constructed using a dynamic clustering procedure (Hénaut and Delorme, 1988) from the same distance-matrix data as that described in the legend of Fig. 2. The important informations derived from this procedure are the clustering pattern and the overall topology. The position of branching points on the x axis reflects the strength of the clustering.

Although not included in our own work, the case of euglenoids deserves some comments ; Sogin *et al.* (1986) have found *Euglena* clearly associated with *Trypanosoma*, emerging near the basis of the eukaryotic tree, in agreement with the concept of "Euglenozoa" (Cavalier-Smith, 1983 ; Corliss, 1984) which groups euglenophytes and kinetoplastids on the basis of morphological similarities. We have failed to sequence *Euglena* by the direct method but since we also found a representative of kinetoplastids (*Crithidia*) emerging at the basis of our tree and our overall topology is nearly identical to that of Sogin and co-workers, we assume that *Euglena* would emerge in our tree at a similar position as in theirs. This would represent a very large distance with respect to chlorophytes and would therefore very strongly suggest that acquisition of chloroplasts by euglenoids was an independent event from the one that gave rise to chlorophytes. Whether this corresponded to the accommodation of a bacterium, of an isolated chloroplast or of a green eukaryote (Gibbs, 1978) cannot, however, be decided from the available data.

<u>Phylogeny of algae</u>

With all types of tree construction methods used (distance and parsimony), we observe a deep separation between chromophytes, chlorophytes and rhodophytes which fits quite well with classical algae systematics. As already emphasized with ciliates (Baroin *et al.*, 1988 ; Lynn and Sogin, 1988), comparative analysis of the rRNA molecule within relatively ancient groups opens the possibility of an independent confrontation to classical systematics. Indeed, within our limited number of photosynthetic representatives, taxonomic inferences indicate the following salient points :

1 - The chromophytes : possible relationships with fungi.

The grouping of a set of algae into chromophytes on the basis of pigment composition, flagellar ultrastructure and the presence of the CER is strongly confirmed by our sequence data. In all our trees, the Chrysophyceae, the Rhaphydophyceae and the Haptophyceae are associated into a single, well individualized branch (see figures 2 to 4). Within this ensemble, well established taxonomic groupings are always disclosed, again confirming the validity of the approach. Thus, the two Haptophyceae (*Prymnesium parvum* and *Cricos roscoffensis*) on one hand and the two Chrysophyceae (*Ochromonas danica* and *Synura petersenii*) on the other hand are clearly clustered. The Rhaphydophyceae (*Vacuolaria virescens*) always lies closer to the Chrysophyceae. The deep demarcation observed between the Haptophyceae and the Chrysophyceae was expected since Haptophyceae are the most remote in structure from the other chromophytes (Hibberd, 1979). Conversely, the relative closeness of the Rhaphydophyceae and the Chrysophyceae is also in agreement with the bulk of structural and biochemical evidence. In fact many authors have joined Rhaphydophyceae to Chrysophyceae in the past. As for dinoflagellates, they have often been placed close to chromophytes on the basis of their photosynthetic pigments. The position of the single representative included in our trees (*P. micans*) is unstable. It often clusters with chromophytes but is also frequently associated with ciliates.

The most striking result concerns the fungi whose origin is still controversial. It has often been proposed that they derive from a red algal ancestor (see Demoulin, 1985 for a review). Although we have sequenced a single red algae, our results do not favor this idea at the moment. In fact, in all our trees, *S. cerevisiae* , *N. crassa* and *F. oxysporum* clearly emerge among chromophytes. Their branching position within chromophytes is, however, quite deep. In addition, Gunderson *et al.* (1987) have found *S. cerevisiae* and *N. crassa* on a separate (although close) branch from their single representative of the chromophytes (*O. danica*) on the basis of complete 18S rRNA sequences : the sharing of a direct common ancestor by chromophytes and fungi must therefore still be taken with reservation although it is a very robust feature of all our trees.

2 - The chlorophytes : origin of the metaphytes.

Due to the similarity of their chloroplastic pigment composition and structure, as well as features of cell division and cell wall composition and structure, it is widely admitted that land plants derive from ancestral chlorophytes. We observe on all our trees that a chlorophyceae (*Chlorogonium elongatum*) and a Prasinophyceae (*Pyramimonas parkae*) branch off independently at the basis of the lineage leading to metaphytes, thus confirming the existence of a common ancestor to chlorophytes and metaphytes. In addition, *C. elongatum* always diverges from the chlorophytes lineage more deeply than *P. parkae* does. This is quite compatible with Melkonian's (1982), Stewart and Mattox's (1975) and Pickett-Heaps' (1972) schemes.

3 - Rhodophytes and cryptophytes : linked but unstable.

Depending on the type of tree construction method used, different branching patterns are observed for these two groups. As seen on figures 2 and 3, the unique representative of rhodophytes we have sequenced (*Porphyridium purpureum*), branches off independently from both the chromophytes and the chlorophytes as a distinctive algal group. It is interesting to note that Cryptophyceae (*Cryptomonas ovata* and the colorless *Chilomonas paramecium*) emerge close to *P. purpureum* (in most trees) but at a deep level of branching. The taxonomic position of Cryptophyceae had always posed difficulties because of its pigment characteristics (chlorophyll a + c, phycobiliproteins) and certain structural features. They have often been accommodated at the basis of the chromophytes series, possibly as intermediates between rhodophytes and chromophytes (Dodge, 1979), or at least as a group deeply split but sharing a common ancestor with chromophytes (Cavalier-Smith, 1986). The distance trees constantly separate them from the chromophytes in agreement with Ludwig and Gibbs' (1985) suggestion. The advantage of this proposal is that it accounts more easily for the difference in morphology of mitochondrial cristae between cryptophytes and chromophytes. However, the distance separating cryptophytes and rhodophytes from the chromophytes is so small compared to the length of the branches that their respective branching order remains uncertain. In addition, their position is somewhat unstable as noted above. In some parsimony tree (not shown), rhodophytes and cryptophytes emerge as a sister group of all algae. In others, cryptophytes and rhodophytes are split and they bracket the other algae. In fact, in most cases the stems are very short so that the situation may best be described at the moment as a trichotomy between rhodophytes + cryptophytes, chromophytes and chlorophytes.

Finally, the rhodophytes and the two cryptophytes are often associated on our trees,albeit at a deep level as noted above. This indicates that the nuclear compartments of these two groups are closely related and suggests the unusual hypothesis that the host to the rhodophyte, that was to becone a cryptophyte, may in fact have been itself a rhodophyte ! This would have facilitated transfer of genes between the cyanobacterial genome of the bacterial symbiont and the nuclei of the two successive hosts.

Origin of the chloroplast : single or multiple ?

In none of our trees did a clear filiation between the various algal divisions emerge. Rather, three distinct possibly monophyletic ensembles were systematically observed : that of chromophytes (which includes the fungi), that of the rhodophytes and cryptophytes and, finally, that of chlorophytes (which diverge from a branch leading to higher plants). Two extreme types of hypotheses can account for such a topology : present day algae do not derive directly from a common photosynthetic ancestor but rather from several independent symbiotic events that gave rise to the various photosynthetic eukaryote groups. As our data presently stand, a minimum of four symbiotic events,two involving a bacterial symbiont and two a eukaryotic one, would have to be hypothesized to account for the major photosynthetic eukaryote groups : a cyanobacterium into a eukaryote to yield rhodophytes ; an unknown photosynthetic eukaryote (usually assumed to be a rhodophyte but see Cavalier-Smith, 1986) into a flagellated eukaryote to yield chromophytes ; a green bacterium into a flagellated eukaryote to yield chlorophytes and a rhodophyte into a flagellated eukaryote to yield cryptophytes. To these basic, possibly unique events, additional independent symbiotic acquisitions must be postulated to account for the green chloroplast of some euglenophytes and the chromophyte-type chloroplast of many dinoflagellates. This multiple symbiosis scheme is basically similar to those proposed by Dodge (1979) and Whatley (1981). It has been reinforced by the discovery of new types of photosynthetic bacteria, in addition to the cyanobacteria, namely the prochorophytes with chlorophylls a + b (see Lewin, 1983) and some c-type chlorophyll containing ones (see Seewaldt and Staskebrandt, 1982). Alternatively, one could argue that the three groups emerge close to each other on most of our trees such that a short common stem is not excluded. In addition, it has been argued that it must have been difficult to carry out successfully the transfer of genes from the initial symbiotic bacterium to the nucleus of the first eukaryotic host (Cavalier-Smith, 1986) endow each of these genes with signal sequences to send the protein back to the chloroplast, etc....

These two arguments would support the monophyletic origin of chloroplasts (Cavalier-Smith, 1983, 1986). In that case it must be postulated that rapid diversification of the initial chloroplast occurred to yield the diverse photosynthetic groups in the short time suggested by the trichotomous shape of our trees. This is basically Cavalier-Smith's view, with the additional suggestion that the cryptophytes and chromophytes share a direct common ancestor which acquired its chloroplat and CER from an endocytized dinoflagellate that still contained both chlorophyll c and phycobillins (see however Schnepf and Elbrächter, 1988 for a different view).

How do the presently available data on chloroplastic and bacterial ribosomal RNAs fit with these views ? The affiliation of green chloroplasts with cyanobacteria is solidly established (Woese 1987 ; Giovannoni *et al.*, 1988) but was there a single origin for the green chloroplast and what about those of rhodophytes and chromophytes ? Two recent papers, one by Turner *et al.* (1989), the other by Witt and Stackebrant (1988) are directly relevant. Turner *et al.* have sequenced a large part of the 16S rRNA of *Prochlorotrix hollandica,* a prochlorophyte that can be grown in culture and by comparison with a large number of homologous bacterial and chloroplastic sequences have found it to lie deeply rooted within the cyanobacteria but not closer to chloroplast sequences of chlorophytes and higher plants than other cyanobacteria and specially the cyanelle of *Cyanophora paradoxa* are. This sequencing work fully confirms the earlier results of Seewaldt and Stackebrandt (1982) on T1-RNase cataloguing of *Prochloron,* a symbiotic prochlorophyte, which was also found to lie out of the line directly leading to green chloroplasts. These findings therefore contradict the idea that present day prochlorophytes share a direct common ancestor with green chloroplasts.

Similarly, Witt and Stackebrandt have analyzed the 16S rRNA of *Ochromonas danica* (a chromophyte) chloroplast and compared it to a set of bacterial and chloroplastic homologues including *Heliobacterium chlorum,* a presumptive bacterial analogue of a "brownish" chloroplast ancestor. Again, *Heliobacterium* was found to lie relatively far from the cluster that contained *Ochromonas* chloroplast together with green chloroplasts and cyanobacteria.

In summary, there is no direct molecular evidence at the moment for the multiple origin of chloroplasts through analysis of chloroplastic and bacterial sequences. However, the number of "non-green" chloroplastic rRNAs analyzed so far is extremely limited and one must await for additional sequences, such as those presently being obtained in Cattolico's laboratoty (Boczar *et al.*, 1989) to definitely settle this point. Also note that the available data while not favouring polyphyly do not prove the monophyly of chloroplasts either since we may simply be missing the appropriate living derivatives of the organism(s) that established the symbiosis(es). The recent data suggesting a biphyletic or mosaic origin of the mitochondrial genome (Gray *et al.*, 1989) should remind us that polyphyly is not such a remote possibility !

Finally, what about the host(s) ? In his single-symbiotic event theory, Cavalier-Smith (1989) argues that the host was a tetraciliated retortamonad-like metamonad cell. Our work, which favors the multiple symbiosis hypothesis, only indicates that the hosts were indeed most probably flagellated. These flagellates would have proliferated during the period of intense radiation discussed earlier in the text, much posteriorly to the divergence of several major "zooflagellate" groups such as Euglenozoa and Parabasalia. At the moment, they appear to be fairly close to ciliates and dinoflagellates but their exact positioning remains to be worked out. In any case, the presumptive chloroplastic symbiotic events appear to have occurred at an advanced stage of protistan diversification and hence, much later than for the mitochondrial counterpart.

We are very grateful to Guy Brugerolle for the generous provison of protist cultures, to Joseph Felsenstein, Alain Hénaut, Marie-Odile Delorme, Desmond Higgins and David Swofford for making their programs generously available, to Hervé Philippe for great help with computation, and to Cécile Couanon for very careful preparation of the manuscript. This work was supported by the Centre National de la Recherche Scientifique, the University of Paris XI and by specific grants from the ATP and ASP "Evolution" and the GdR "Evolution des organismes microscopiques" of the Centre National de la Recherche Scientifique.

References :

Baroin A, Perasso R, Qu LH, Brugerolle G, Bachellerie JP, Adoutte A (1988) Partial phylogeny of the unicellular eukaryotes based on rapid sequencing of a portion of 28S ribosomal RNA. Proc Natl Acad Sci USA 85:3474-3478

Boczar BA, Delaney TP, Cattolico RA (1989) Gene for the ribulose-1,5 bisphosphate carboxylase small subunit protein of the marin chromophyte *Olisthodiscus luteus* is similar to that of a chemoautotrophic bacterium. Proc Natl Acad Sci USA 86:4996-4999

Burger-Wiersma T, Veenhuis M, Korthals HJ, Van De Wiel CCM, Mur LR (1986) A new prokaryote containing chlorophyll *a* and *b*. Nature 320:262-264

Cavalier-Smith T (1983) A 6-kingdom classification and a unified phylogeny. In : Endocytobiology Vol II Walter de Gruyter & Co Berlin New York pp 1027-103

Cavalier-Smith T (1986) The kingdom chromista : origin and systematics. In: Round, Chapman (eds) Progress in Phycological Research, Vol 4, Biopress Ltd, 309-347

Cavalier-Smith T (1989) The kingdom Chromista. In: Green JC, Leadbeater BSC (eds) The Chromophyte Algae: problems and perspectives, Oxford University Press

Cedergren R, Gray MW, Abel Y, Sankoff D (1988) The evolutionary relationships among known life forms. J Mol Evol 28:98-112

Corliss JO (1984) The kingdom protista and its 45 phyla. BioSystems 17:87-126

Demoulin V (1985) The red algal-higher fungi phylogenetic link : the last ten years. BioSystems 18:347-356

Dodge JD (1979) The Phytoflagellates : Fine structure and phylogeny. In : Levandowsky M, Hutner SH (eds) Biochemistry and physiology of Protozoa, Vol 1. 2nd ed. Academic Press, New York London, p 7

Felsenstein J (1978) Cases in which parsimony or compatibility methods will be positively misleading. Syst Zool 22:240-249

Felsenstein J (1982) Numerical methods for inferring evolutionary trees. Quaterly Rev Biol 57:379-404

Felsenstein J (1983) Parsinomy in systematics : Biological and statistical issues. Ann Rev Ecol Syst 14:313-333

Felsenstein J (1988) Phylogenies from molecular sequences : inference and reliability. Ann Rev Genet 22:521-565

Fink WL (1986) Microcomputers and phylogenetic analysis. Science 234:1135-1139

Fitch WM, Margoliash E (1967) Construction of phylogenetic trees. A method based on mutation distances as estimated from cytochrome c sequences is of general applicability. Science 155:279-284

Gibbs SP (1970) The comparative ultrastructure of the algal chloroplast. Ann New York Acad Sci 175:454-473

Gibbs SP (1978) The chloroplasts of *Euglena* may have evolved from symbiotic green algae. Can J Bot 56:2883-2889

Gibbs SP (1981) The chloroplasts of some algal groups may have evolved from endosymbiotic eukaryotic algae. Ann New York Acad Sci 361:193-208

Giovannoni SJ, Turner S, Olsen GJ, Barns S, Lane DJ, Pace NR (1988) Evolutionary relationships among cyanobacteria and green chloroplasts. J Bacteriol 170:3584-3592

Gray MW (1988) Organelle origins and ribosomal RNA. Biochem Cell Biol 66:325-348

Gray MW, Cedergren R, Abel Y, Sankokk D (1989) On the evolutionary origin of the plant mitochondrion and its genome

Gray MW, Doolittle WF (1982) Has the endosymbiont hypothesis been proven ? Microbiol Rev 46:1-42

Gunderson JH, Elwwod H, Ingold A, Kindle K, Sogin ML (1987) Phylogenetic relationships between chlorophytes, chrysophytes, and oomycetes. Proc Natl Acad Sci 84:5823-5827

Hénaut A, Delorme MO (1988) Distance matrix comparison and tree construction. Pattern Recognition Letters 7:207-213

Hibberd DJ (1979) The structure and phylogenetic significance of the flagellar transition region in the chlorophyll c-containing algae. BioSystems 11:243-261

Holmquist R, Miyamoto MM, Goodman M (1988a) Higher-primate phylogeny - Why can't we decide ? Mol Biol Evol 5:201-216

Holmquist R, Miyamoto MM, Goodman M (1988b) Analysis of higher-primate phylogeny from transversion differences in nuclear and mitochondrial DNA by Lake's methods of evolutionary parsimony and operator metrics. Mol Biol Evol 5:217-236

Hori H, Osawa S (1987) Origin and evolution of organisms as deduced from 5S ribosomal RNA sequences. Mol Biol Evol 4:445-472

Hunt LT, George DG, Barker WC (1985) The prokaryote-eukaryote interface. BioSystems 18:223-240

Kimura M (1981) Estimation of evolutionary distances between homologous nucleotide sequences. Proc Natl Acad Sci USA 78:454-458

Lake JA (1987) A rate-independent technique for analysis of nucleic acid sequences : evolutionary parsimony. Mol Biol Evol 4:167-191

Lane DJ, Pace B, Olsen GJ, Stahl DA, Sogin ML, Pace NR (1985) Rapid determination of 16S ribosomal RNA sequences for phylogenetic analyses. Proc Natl Acad Sci USA 82:6955-6959

Lewin RA (1983) The problems of *Prochloron*. Ann Microbiol (Inst Pasteur) 134B:37-41

Li WH, Wolfe KH, Sourdis J, Sharp PM (1987) Reconstruction of phylogenetic trees and estimation of divergence times under nonconstant rates of evolution. In: Cold Spring Harbor Symposia on Quantitative Biology, Cold Spring Harbor Lab. Vol LII:847-856

Ludwig M, Gibbs SP (1985) DNA is present in the nucleomorph of cryptomonads: further evidence that the chloroplast evolved from a eukaryotic endosymbiont. Protoplasma 127:9-20

Lynn DH, Sogin ML (1988) Assessment of phylogenetic relationships among ciliated protists using partial ribosomal RNA sequences derived from reverse transcripts. BioSystems 21:249-254

Margulis L (1981) Symbiosis in cell evolution : Life and its environment on the early earth. Freeman W.H., San Francisco

Melkonian M (1982) Structural and evolutionary aspects of the flagellar apparatus in green algae and land plants. Taxon 31:255-265

Nanney DL, Preparata RM, Preparata FP, Meyer EB, Simon EM (1989) Shifting ditypic site analysis: heuristics for expanding the phylogenetic range of nucleotide sequences in Sankoff analyses. J Mol Evol 28:451-459

Olsen GL (1987) Earliest phylogenetic branchings: comparing rRNA-based evolutionary trees inferred with various techniques. In: Cold Spring Harbor Symposia on Quantitative Biology, Cold Spring Harbor Lab. Vol LII:825-837

Olsen GJ (1988) Phylogenetic analysis using ribosomal RNA. In: Methods in Enzymology. Acad. Press Inc. Vol 164:793-812

Penny D, Hendy MD, Henderson IM (1987) Reliability of evolutionary trees. In : Cold Spring Harbor Symposia on Quantitative Biology, Cold Spring Harbor Lab. Vol LII:857-862

Pickett-Heaps JD (1972) Variation in mitosis and cytokinesis in plant cells : its significance in the phylogeny and evolution of ultrastructural systems. Cytobios 5:59-77

Qu LH, Michot B, Bachellerie JP (1983) Improved methods for structure probing in large RNAs: a rapid 'heterologous' sequencing approach is coupled to the direct mapping of nuclease accessible sites. Application to the 5' terminal domain of eukaryotic 28S rRNA. Nucl Acid Res 11:5903-5920

Saitou N, Nei M (1987) The neighbor-joining method: a new method for reconstructing phylogenetic trees. Mol Biol Evol 4:406-425

Saitou N, Imanishi T (1989) Relative efficiencies of the Fitch-Margoliash, maximum-parsimony, maximum-likelihood, minimum-evolution, and neighbor-joining methods of phylogenetic tree construction in obtaining the correct tree. Mol Biol Evol6:514-525

Schnepf E, Elbrächter M (1988) Cryptophycean-like double membrane-bound chloroplast in the dinoflagellate, *Dinophysis* Ehrenb.: evolutionary, phylogenetic and toxicological implications. Botanica Acta 101:196-203

Seewaldt E, Stackebrandt E (1982) Partial sequence of 16S ribosomal RNA and the phylogeny of *Prochloron*. Nature 295:618-621

Sogin ML, Elwood HJ, Gunderson JH (1986) Evolutionary diversity of eukaryotic small-subunit rRNA genes. Proc Natl Acad Sci USA 83:1383-1387

Sogin ML, Gunderson JH, Elwood HJ, Alonso RA, Peattie DA (1989) Phylogenetic meaning of the kingdom concept: an unusual ribosomal RNA from *Giardia lamblia*. Science 243:75-77

Sourdis J, Krimbas C (1987) Accuracy of phylogenetic trees estimated from DNA sequence data. Mol Biol Evol 4:159-166

Sourdis J, Nei M (1988) Relative efficiencies of the maximum parsimony and distance-matrix methods in obtaining the correct phylogenetic tree. Mol Biol Evol 5:298-311

South GR, Whittick A (1987) Introduction to phycology. Blackwell Scientific Publications Oxford London Edinburgh Boston Palo Alto Melbourne

Stewart KD, Mattox KR (1975) Comparative cytology, evolution and classification of the green algae with some consideratioon of the origin of other organisms with chlorophylls *a* and *b*. Bot Rev 41:104-135

Turner S, Burger-Wiersma T., Giovannoni SJ, Mur LR, Pace NR (1989) The relationship of a prochlorophyte *Prochlorothrix hollandica* to green chloroplasts. Nature 337:380-382

Vossbrinck CR, Maddox JV, Friedman S, Debrunner-Vossbrinck BA, Woese CR (1987) Ribosomal RNA sequence suggests microsporidia are extremely ancient eukaryotes. Nature 326:411-414

Whatley JM (1981) Chloroplast evolution - ancient and modern. In: Fredrick JF (ed) Origins and evolution of eukaryotic intracellular organelles. New York Acad. Sci. New York. Annals New York Acad. Sci. 361:154-165

Witt D, Stackebrandt E (1988) Disproving the hypothesis of a common ancestry for the *Ochromonas danica* chrysoplast and *Heliobacterium chlorum*. Arch Microbiol 150:244-248

Woese CR (1987) Bacterial evolution. Microbiol. Rev. 51:221-271

Algae as tools in studying the biosynthesis of cellulose, nature's most abundant macromolecule

R. Malcolm Brown, Jr.
Department of Botany
The University of Texas at Austin
Austin, Texas 78712
U.S.A.

INTRODUCTION

The most dominant polysaccharide of the cell wall is cellulose. The universal distribution of this natural polymer among procaryotic and eucaryotic organisms attests to its ancient evolutionary history. Not only is cellulose found among photosynthetic and protistan cells, it is present in animals such as the Ascidians (Wardrop, 1970). Furthermore, levels of elevated cellulose synthesis have been suggested in humans with the disease scleroderma (Hall, et al, 1960).

The algae have been prominent organisms of study among eucaryotic organisms because of their great diversity of structure and cellular organization. Chloroplast morphology (Gibbs, 1981), organization of the mitotic apparatus (Pickett-Heaps, 1972), cell wall structure and composition (Preston, 1974), flagellar apparatus (Stewart and Mattox, 1978), and reproduction have provided a wealth of information on algal phylogeny (Stewart and Mattox, 1982); however, very few studies have concentrated on evolutionary and phylogenetic aspects of cell walls, let alone cellulose, mainly because the cellulose synthases had never been observed or isolated.

In 1976, the first successful application of freeze fracture demonstrated the structure of a cellulose synthase complex in *Oocystis apiculata* (Brown and Montezinos, 1976). These membrane associated structures, called terminal synthesizing complexes (=TCs), were found at the growing tip of microfibril impressions on the E-fracture face of the plasma membrane. Earlier, Roelofsen (1958) had predicted that an organized terminal enzyme complex would be found; however, the shape and geometry could not be predicted at that time.

In 1964, Preston proposed the _ordered granule hypothesis_ for the cellulose synthase complex. This study was based on observations of remnants of organized particle subunits associated with the innermost wall of _Chaetomorpha_. During this time, the freeze fracture technique was becoming widely used, and organized particle complexes were found in the plasma membranes, particularly abundant in yeasts (Moor and Muhlethaler, 1963). Thus, a logical extension of the possibility of such an organization could be made for the cellulose synthesizing complex. Interestingly, TCs were not observed in cells which had been chemically fixed or treated with glycerol or cryoprotectants. The breakthrough in finding TCs came when Brown and Montezinos demonstrated that rapid direct freezing of living cells yielded organized particle structures associated with the tips of microfibrils.

Since 1976, TCs have been found in more than 14 algal genera, and presently, a distinct pattern of TC structure is beginning to emerge which provides some insight into the substantial diversity of cellulose microfibril synthesis among the algae. In this presentation, the fundamental TC diversity among the algae will be described. Because TC variation is the greatest among the algae, it also provides hints at the relationship between TC organization and microfibril shape, molecular weight of the cellulose, and crystallization. Phylogenetic relationships based on TC structure can be correlated with other structural and biochemical evidence. This presentation will conclude with a proposed evolutionary history of cellulose biogenesis.

TERMINAL COMPLEX DIVERSITY

So far, only two basic forms of TCs have been found among eucaryotic cells. The linear TC was first discovered in _Oocystis_ and is present among most members of the Ulvophyceae (Hotchkiss and Brown, 1989) (Fig 1). The rosette TC was reported simultaneously by Mueller and Brown (1980) and

Giddings, et al (1980). Earlier, Mueller, et al (1976) had observed a globular component of the TC on the E-fracture face of the plasma membrane in corn roots. Subsequent examination of the P-fracture face revealed a rosette of 6 symmetrically organized particle subunits (Fig 2).

Unlike that in the corn root, the rosette TC found in Micrasterias frequently is organized in ordered arrays, especially during secondary wall formation (Giddings, et al, 1980). Furthermore, the rosette subunits have been shown to span both layers of the bimolecular leaflet. Since these first reports, rosette TC's have been found in a variety of organisms, ranging from the land plants to Chara (McLean and Juniper, 1986) and Nitella (Hotchkiss and Brown, 1987).

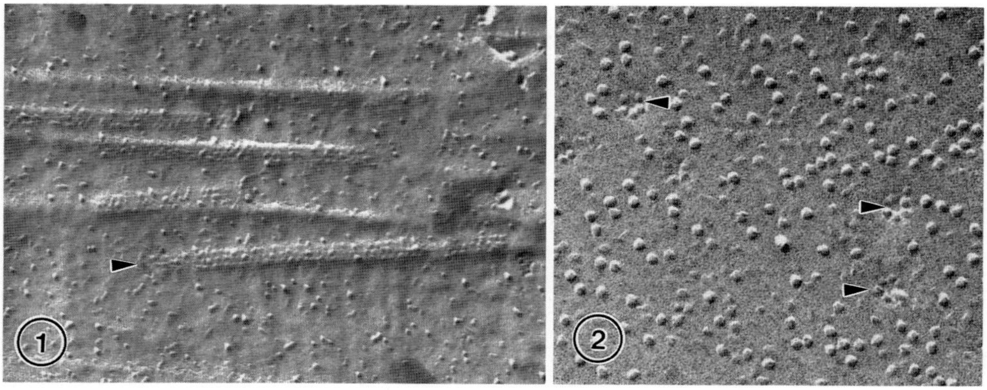

Figs 1 and 2. TCs in Oocystis(left) and Zea mays(right).

Interestingly, only the rosette TC has been found among the vascular plants. Thus, algal ancestors with rosette TCs have assumed added importance with respect to understanding the evolution of land plants.

For a more detailed elaboration of the distribution of TC's among eucaryotic cells, consult the report of Brown (1985). Since this publication, TCs have been found in other organisms and will briefly described. In 1987, Hotchkiss and Brown described a solitary rosette TC in Nitella translucens (Hotchkiss and Brown, 1987). This TC is virtually indistinguishable from that of land plants. In addition, a solitary rosette TC has been observed in Chara (McLean and Juniper, 1986).

Only in the Zygnematales has an ordered consolidation of TC's been found, and among one member, Mougeotia, the TCs are exclusively solitary (Hotchkiss et al 1989). Thus, the consolidation of rosette TCs appears to have been a more recent event (Hotchkiss and Brown, 1989a). When wall composition is taken into consideration, the evidence strongly supports that the Charophyceae is the phylogenetic line which may have given rise to the vascular plants.

Another interesting and different TC structure recently was found in the xanthophycean genus, Vaucheria (Mizuta et al, 1989). Overall, the TC is linear, but the subunit arrangement within the TC is in the form of unique diagonal rows (Fig 3).

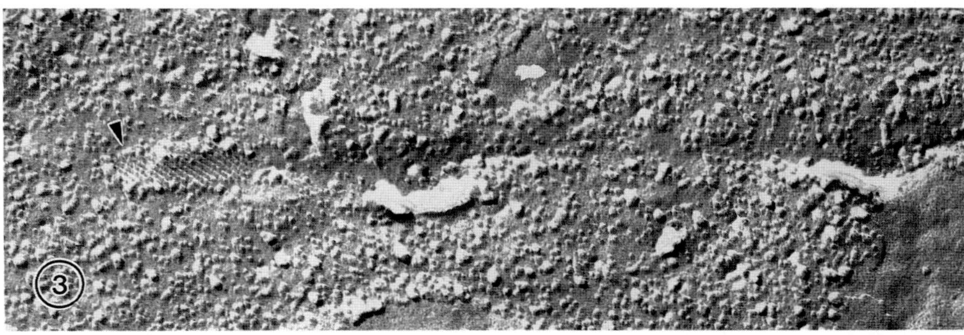

Fig 3. The linear TC of Vaucheria with diagonal rows of subunits, (from Mizuta and Brown).

The relationship between TC geometry and microfibril shape is better understood in Vaucheria (Mizuta et al, 1989). Each subunit of the diagonal row assembles a single glucan chain. Glucan chain products of the diagonal row aggregate to form an ordered aggregate. Aggregates unite laterally to form a thin ribbon-shaped microfibril. Whether or not the peculiar subunit geometry within the linear TC of Vaucheria is indicative of other xanthophycean algae is unknown at present. Vaucheria is yet but another example demonstrating the great diversity of cellulose biogenesis among the algae. With the rudimentary evidence at hand, one can now consider the evolution of cellulose biogenesis.

THE EVOLUTION OF THE TC AND CELLULOSE BIOGENESIS

Based on TC ultrastructure and cellulose organization, it is possible to construct a phylogenetic pathway for cellulose biogenesis (Fig 4). Although somewhat premature, this treatment gives several important and interesting clues to the subject of eucaryotic cellular evolution. Clearly, the algae as a group have been pivotal in the diversification of cellulose microfibril assembly as exemplified by the diversity of TC and microfibril morphology.

Procaryotic Cellulose

The evolution of cellulose among procaryotic cells will first be considered. The fact that cellulose assembly occurs among two widely divergent procaryotic groups suggests that this process must have been an ancient one. The investigations of Woese have shed much light on bacterial evolution. Using homologous sequences in rRNA, it has been possible to unambiguously examine and measure phylogenetic relationships among the bacteria (Woese, 1987).

The genus, *Sarcina* is a gram + bacterium and is considered primitive among the eubacteria. *Sarcina* also is an obligate anaerobe. Cellulose synthesis in *Sarcina* was first described by Canale-Parola *et al* (1961). We have also studied cellulose synthesis in *Sarcina* (Roberts and Brown, unpublished data) and have come to the conclusion supporting Canale-Parola et al that only the cellulose II polymorph is synthesized. The exact site of cellulose synthesis is unknown; however, membrane fractions of *Sarcina* can synthesize cellulose II *in vitro* (Lin, Roberts, and Brown, unpublished data).

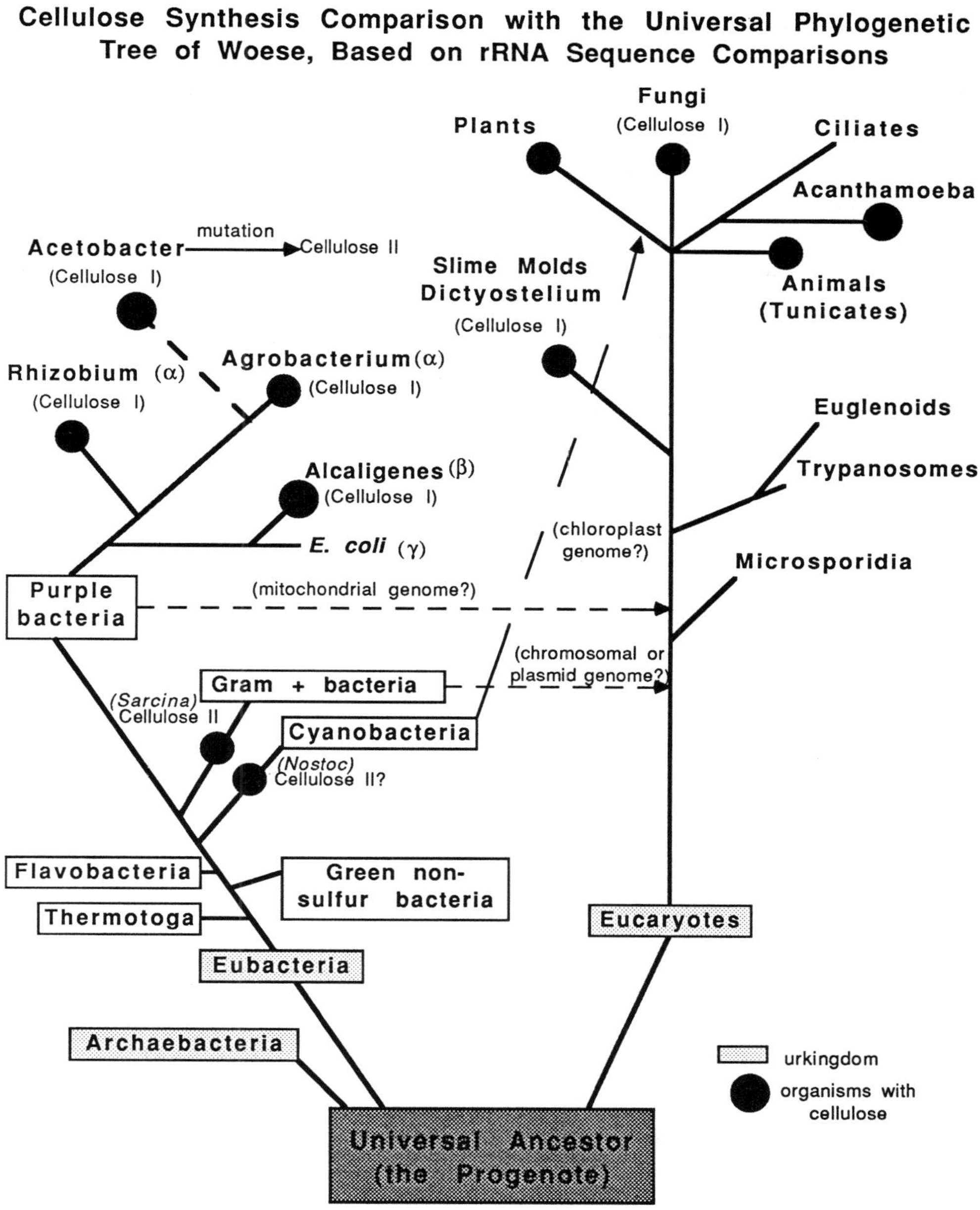

Fig 4.Scheme of the evolution of cellulose synthesis based on the universal phylogenetic tree of Woese.

Among the procaryotes, the purple bacteria as a group are thought to be more advanced (Woese, 1987). They have been considered as progenitors of mitochondria in the eucaryotic cell (Yang <u>et al</u>, 1985). Interestingly, many genera in this group synthesize cellulose. These include, *Acetobacter*, *Rhizobium*, and *Agrobacterium* (Deinema and Zevenhuizen, 1971). As Woese points out, the purple bacteria are closely related and seem to have evolved special relationships with vascular plants as exemplified by nodules for nitrogen fixation in *Rhizobium* to tumors in *Agrobacterium*. The cellulose of *Agrobacterium* has been proposed to aid in establishing the virulence although it is not critical (Matthysee, 1983). The cellulose synthesized by *Rhizobium* may aid in attaching the bacterium to the root hair tip where it will initiate infection thread formation (Rolfe and Gresshoff, 1988). In addition, cellulose is synthesized in *Alcaligenes*; however no cellulose synthesis has been observed in *E. coli*.

Among the purple bacteria, only the cellulose I polymorph is synthesized (one exception is a mutation in *Acetobacter* giving rise to cellulose II synthesis, see Roberts *et al*, 1989). The site of cellulose microfibril synthesis is a linear row of particles parallel to the longitudinal axis of the cell (Brown *et al*, 1976). On the surface of the LPS layer of the cell envelope is a pore complex through which fibrils are extruded. The enzyme complex is located in the cytoplasmic membrane (Bureau and Brown, 1987). Groups of three or more particles are required to assemble subfibrils which consolidate to form a microfibril tangential to the cell surface (Haigler *et al*, 1980). Microfibrils unite to form bundles which, in turn, group to assemble ribbons. The ribbons are visible with darkfield microscopy, and the progress of cellulose synthesis can be directly monitored using time lapse video microscopy (Brown and Santos, unpublished observations; Brown and Colpitts, 1978; Lin and Brown, 1989).

In *Rhizobium* and *Agrobacterium*, the sites of cellulose assembly are similar; however, fewer microfibrils consolidate. Only a floc of cellulose is produced, while in *Acetobacter*, a thick leathery membrane or pellicle is synthesized. The function of cellulose biosynthesis in *Acetobacter* is unknown. *Acetobacter xylinum* is an obligate aerobe. Therefore the buoyancy of its cellulose could provide an aerobic environment for cells active in cellulose synthesis (Schramm and Hestrin, 1954). Why *Acetobacter* would divert large pools of metabolic substrate into cellulose is unknown; however, cellulose could serve as a reserve pool for metabolism. Unfortunately, extensive cellulase activity has not been found; yet, degradation of non-crystalline carboxymethylcellulose can occur (Brown and White, unpublished observations).

Fundamental research investigations of cellulose biogenesis in *Acetobacter* have yielded perhaps the most extensive evidence for the molecular mechanisms of cellulose assembly. Cellulose has now been synthesized *in vitro* (Glaser, 1958); a specific activator for cellulose synthesis has been found (Ross *et al*, 1985), the *in vitro* product characterized morphologically (Lin *et al*, 1985), the *in vitro* crystalline polymorph deduced (Bureau and Brown, 1987), the cellular site of the synthase localized (Bureau and Brown, 1987), the cellulose synthase purified (Lin and Brown, 1989), and gene cloning for the purified enzyme initiated (Saxena and Brown, 1989).

Cellulose synthesis among the blue green algae has been mentioned only once in the literature, and conclusive evidence for this is lacking (Frey-Wyssling, 1976). According to Woese (1987), the cyanobacteria emerged from the distance matrix phylogenetic tree of the eubacteria almost from the same region as the gram + bacteria and also close to the purple bacteria. Thus, it is clear that cellulose biogenesis must have been an ancient process during the evolution of life on earth.

With the accumulating evidence of cellulose biogenesis among procaryotes, the evolution of cellulose among this group can now be addressed. The cellulose II polymorph appears to be primitive. This polymorph is the more thermodynamically stable form with an additional inter-chain H-bond formed per glucose residue. Cellulose II can form spontaneously from solution when cellulose is solubilized by such agents as DMSO/paraformaldehyde or cupraammonium reagents (Blackwell *et al*, 1986). On the other hand, cellulose I is a metastable polymorph, presumably assembled only by living organisms. Therefore, cellulose I is more advanced, requiring additional mechanisms for chain orientation and positioning to achieve the metastable state.

The first cellulose producer probably had its glucan synthase randomly organized in association with the membrane (Fig 5). Under these primitive conditions, only cellulose II could be synthesized *in vivo*. Thus, in *Sarcina*, glucan chains appear to be randomly positioned over the cell surface in an amorphous array. *Sarcina* is an excellent example of an extant organism which lacks mechanisms to consolidate and organize the glucan synthase complex to induce ordered microfibril assembly and aggregation into bundles and ribbons of cellulose I. On the other hand, the purple bacteria evolved the mechanisms to order glucan chains into various conformations leading to a diversity of cellulose I microfibril assembly. Because of abundant evidence suggesting that the purple bacteria are more advanced (Woese, 1987), it follows that cellulose I assembly is more advanced. A logical extension of this evolutionary advancement is to consider the evolution of cellulose among the eucaryotes. Did these cells obtain genes for cellulose synthesis from the eubacteria? If so, what are the likely candidates? It is obvious that if the mitochondria of eucaryotes came from the purple bacteria, did they also transfer the genes for organized cellulose I synthesis? This is an intriguing question, one which must await cloning and sequencing of the genes for the cellulose synthases.

Eucaryotic Cellulose

We shall now consider one of the greatest mysteries in biology- the origin of cellulose synthesis among eucaryotic cells. The possibility that the genes for cellulose synthesis may have come from the purple bacteria has been alluded to. What was the first eucaryotic organism to have received the cellulose synthase gene? An examination of phylogenetic trees indicates that among the extant primitive eucaryotes, cellulose is found among one group, namely, the cellular slime molds. Cellulose I of low crystallinity has been found in Dictyostelium (Roberts and Brown, unpublished data). Because Dictyostelium is non-photosynthetic, it seems highly unlikely that it could have received the genes for cellulose synthesis from a photosynthetic procaryotic progenitor. It seems attractive, therefore, that the purple bacteria may have donated these complexes.

Dictyostelium TCs are unlike any so far found among eucaryotic cells. They appear to consist of linear arrays of single particles (Fig. 4) (Mizuta and Brown, unpublished results). These arrays are somewhat similar to those found in Acetobacter. Thus, among eucaryotic organisms, the linear arrangement of single rows of particle subunits in Dictyostelium appears to be primitive. This is also supported by the low crystallinity of cellulose I from this organism.

Fig 4. Linear TCs from a pre-stalk cell of Dictyostelium discoideum (unpublished electron micrograph of S. Mizuta and R. M. Brown, Jr,)

Certain fungi, among them, the Oomycetes, synthesize cellulose. Yet, nothing is known of TC structure among the fungi. This is a major research area which needs immediate attention. One could predict, however, that if Vaucheria is closely related to Saprolegnia, the unique linear TC with diagonal rows (Mizuta and Brown, 1989) might be found in the latter.

Why many of the fungi opted for chitin as the major wall polymer is an interesting sidelight to the question of the evolution of cellulose synthesis. Some of the fungi are almost as primitive as *Dictyostelium*, yet there is scant knowledge of chitin synthesis among primitive eucaryotes, let alone the eubacteria and archaebacteria. Perhaps nutrition may have played a role in the more efficient utilization of strong polymers. Since chitin walled organisms may have required more nitrogen, the energy budget would be greater. Thus, to synthesize a polymer of cellulose would represent a selective advantage from a nutritional point of view (Duchesne and Larson, 1989).

THE EVOLUTION OF CELLULOSE SYNTHESIS AMONG PHOTOSYNTHETIC EUCARYOTES

At this junction, the algae certainly have played a diversified role, for it is here that the greatest variety of TCs and cellulose structure is found among the algae (Fig 6). Gunderson *et al*, 1987) have suggested on the basis of 18S rRNA sequences that a cellulose producing oomycete, *Achlya bisexualis*, is closely related to *Ochromonas danica*, a chitin producer. Perhaps the fungi and chromophytes appeared at roughly the same time.

The recent studies of Pearasso *et al*, 1989) have shed some light on the origin of the algae. On the basis of 28S cytoplasmic rRNA homology, they found that three distinct groups emerged late among eucaryotes: rhodophytes, chromophytes, and chlorophytes. A late occurrence of eucaryotic photosynthetic symbiosis was implied. The conserved rosette/linear TCs among the algae suggests that the synthase may have come from the more primitive fungi or *Dictyostelium*, rather than through the chloroplast. Again, this implies that even earlier, the ancient eucaryotic mitochondrion may have contained the genes for cellulose synthase.

Could the cellulose synthase of the Rhodophyta have come from the cyanobacteria? This question cannot be answered until we have more data on the presence and physical characteristics of cellulose among the cyanobacteria. If only cellulose II is present among the cyanobacteria, it would be difficult to imagine an independent evolutionary event to organize the TC to allow it to produce cellulose I which is found among the Rhodophyta. If, however, Nostoc is found to produce cellulose I in vivo, the hypothesis that the cyanobacteria could have been the progenitors of the Rhodophyta, would be strengthened. It should be apparent that these questions cannot really be answered, let alone seriously considered, until we have sequence information on the cellulose synthase; yet these questions do need to be placed before the scientific community now so that the blueprint for solving this great mystery can be expedited.

The origin of cellulose synthesis among the major algal groups is still a major mystery. Consider that cellulose is known among the Pyrrophyta, the Chrysophycease, the Xanthophyceae, the Phaeophyta, and the Chlorophyta (Fig 6). Did each of these major groups receive a cellulose synthase independently, possibly from pro-chloroplast capture? The diversity of TC structure among these groups might suggest multiple independent captures, yet only the purple bacteria are known to synthesize cellulose I, the same polymorph found among the great diversity of algae (with the exception of Halicystis, Sisson, 1938; Roberts and Brown, 1989, unpublished data). This argues in favor of a single capture of the cellulose synthase very early in eucaryote evolution, possibly through the purple bacteria donation of the pro-mitochondrial apparatus. This implies that the organizational machinery for glucan chain assembly into the cellulose I polymorph may have also been introduced early through transfer to an ancient eucaryotic progenitor. Then, Dictyostelium may be an example of one of the most ancient surviving groups which received the cellulose synthase from a procaryotic progenitor. These are provocative questions, but ones which cannot be answered with certainty at present.

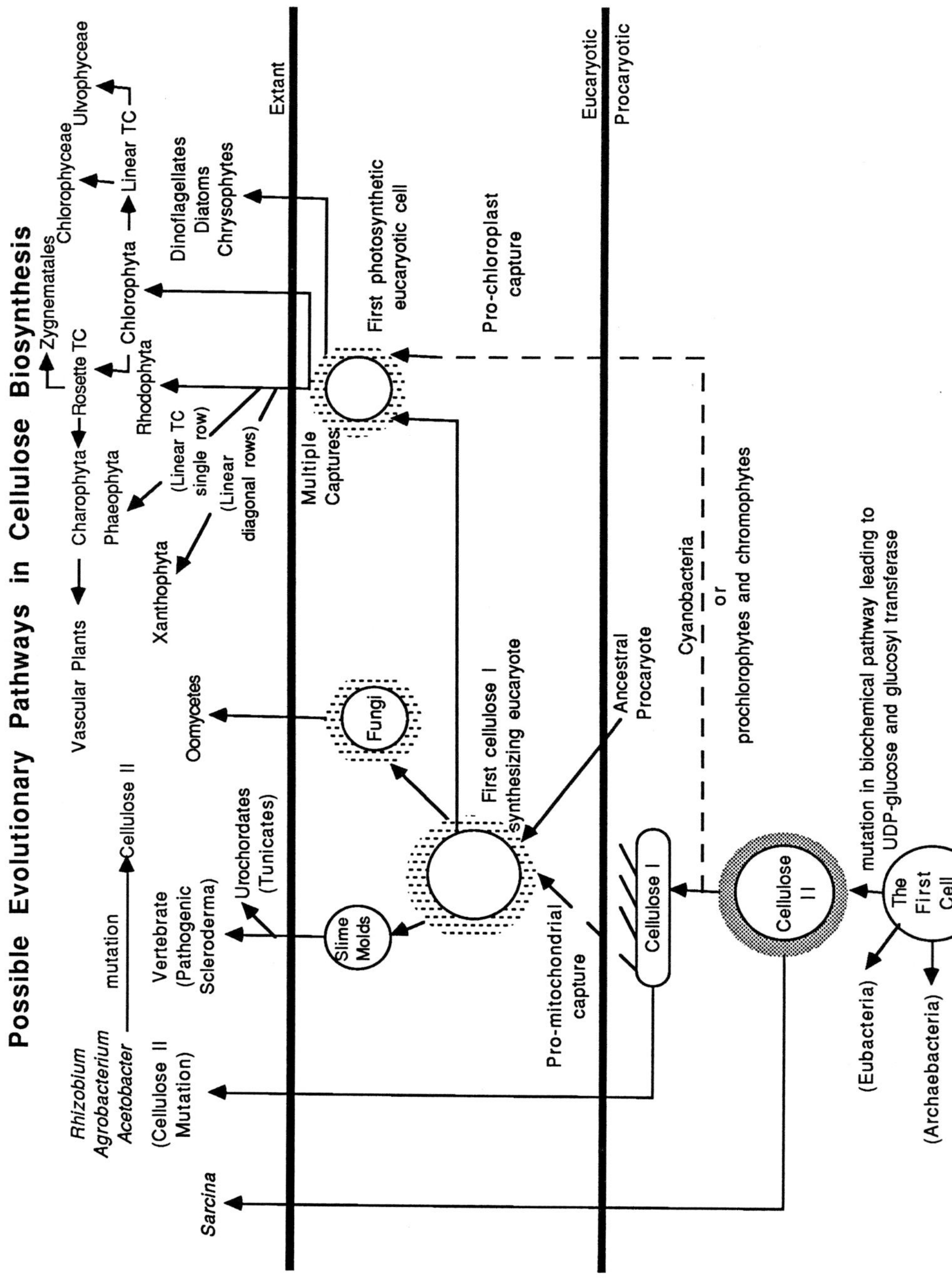

Fig 5. Possible evolutionary pathways of cellulose biogenesis, showing the relationship of the algae to early life.

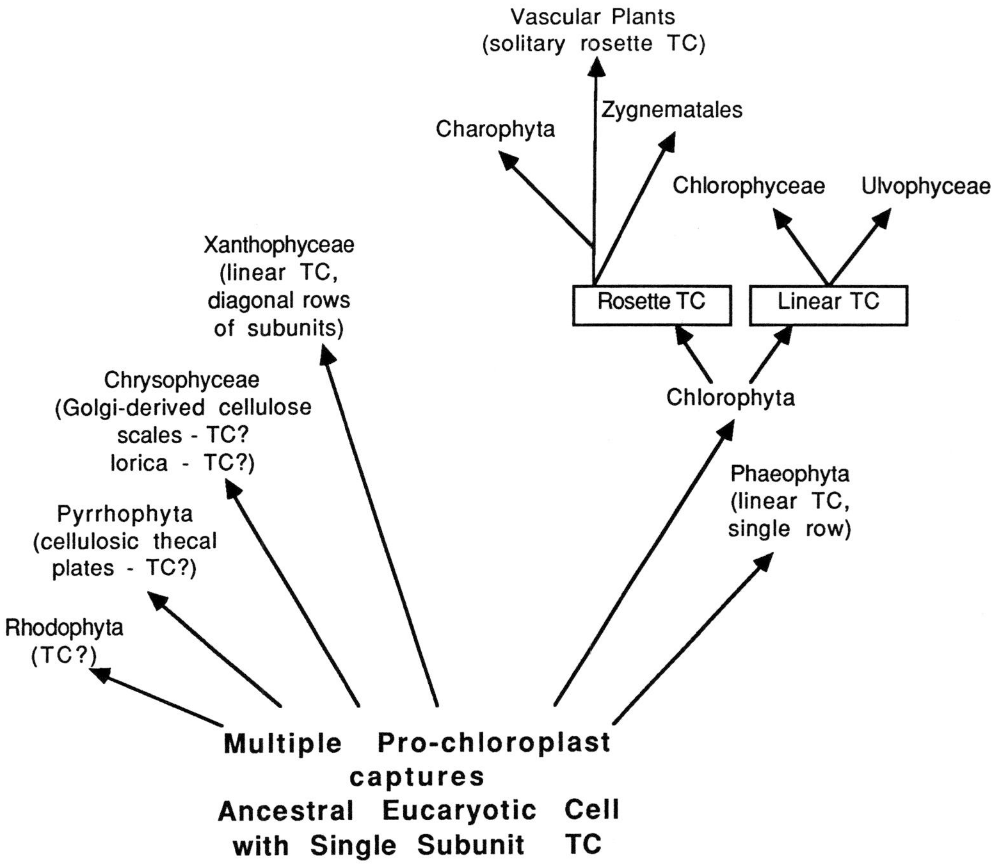

Fig 6. Evolutionary relationships of TC structure among the algae and possible origins of TCs from symbiotic capture.

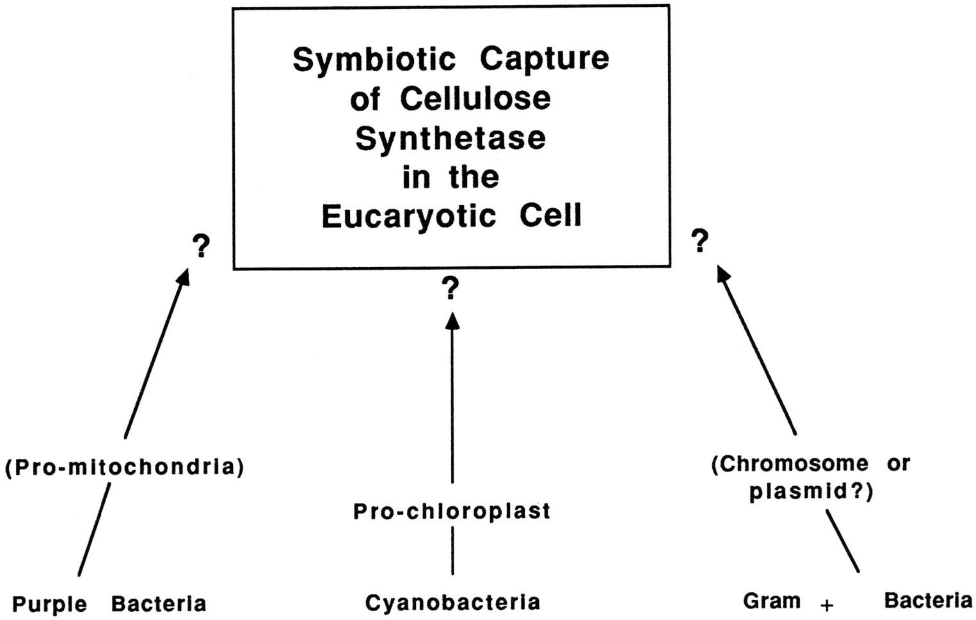

Fig 7. Possible sources of symbiotic capture of the cellulose synthase.

CONCLUSIONS

Cellulose synthase is an ancient molecule as evidenced by its ubiquity among procaryotic and ecaryotic organisms. It therefore must have served an important early function in the origin of life on earth. The most advanced eucaryotic land plants have the rosette cellulose synthase TC. Advanced members of the Charophyta have the rosette TC, an so do members of the Zygnemetales. At present, all pieces to the puzzle are not in place. Critically important algae need to be investigated for their TC structure to provide more direct confirming data for the rosette TC in land plant evolution. These include _Coleochaetae_, members of the Trentepholiales (_Cephaleuros_), the Klebsormidiales (_Klebshormidium_), the Chlorokybales (_Chlorokybus_), and other members of the Zygnematales (_Zygnema, Netrium_).

The concept of symbiotic capture of the cellulose synthase in eucaryotic cells needs more study (Fig 7). Did the cellulose synthase come from a mitochondrial progenitor (through the purple bacteria)? Did it originate through a chloroplast progenitor (from the cyanobacteria, prochlorophyta, or pro-chromophyta)? Did the cellulose synthase transfer directly from the procaryote chromosome or plasmid? Perhaps one or more symbiotic captures occurred, giving rise the the diversity of TC morphology and microfibril diversity.

The known diversity of cellulose microfibril structure, degree of polymerization, and crystallinity, suggests that secondary modifications to the primary glucan synthase must have occurred. Perhaps even the regulation of cellulose synthase activity has been modified, although cellulose synthase activity appears largely constitutive. The diversity of fibrillar polymers other than cellulose among the algae supports an extremely diverse evolution in cell wall structure, possibly reflecting the wide range of environments for survival.

Algae have served as excellent model systems in the quest for understanding cellulose biogenesis. Not only do the algae synthesize diverse forms off cellulose, they offer an uncompromising approach for cytological obervation, coupled with a potential for isolation, cloning, and sequencing of the genes involved in cellulose biogenesis. The coming decade will see great advances in cellulose biogenesis research.

ACKNOWLEDGEMENTS

Appreciation is expressed to Richard Santos for his unfailing loyalty and support throughout the past 19 years. Also, my former and present graduate students, post-doctoral fellows, and visitors to the lab deserve recognition for their contributions to this research. I thank Inger Johansen for preparing this manuscript. This research was supported by the National Science Foundation, The Johnson & Johnson

Centennial Chair Endowment, and funds from the University of Texas at Austin.

REFERENCES

Blackwell J, Lee DM, Kurz D, Su MY (1986) Structure of cellulose-solvent complexes. In: Young R, Rowell RM (eds) Cellulose-Structure, Modification, and Hydrolysis. John Wiley and Sons, New York, p 51-66.

Brown Jr RM, Montezinos DL (1976) Cellulose microfibrils: visualization of biosynthetic and orienting complexes in association with the plasma membrane. Proc Natl Acad Sci USA 73: 143-147.

Brown Jr RM, Willison JHM, Richardson CL (1976) Cellulose biosynthesis in *Acetobacter xylinum*: visualization of the site of synthesis and direct measurement of the *in vivo* process. Proc Natl Acad Sci USA 73:4565-4569.

Brown Jr RM, Colpitts TJ (1978) Direct visualization of cellulose synthesis by high resolution darkfield microscopy and time-lapse cinematography. J Cell Biol 79(2):157a.

Brown Jr RM (1985) John Innes Symposium - Cellulose microfibril assembly and orientation: recent developments. J Cell Sci Suppl 2:13-32.

Brown Jr RM, Lin FC (1989) Time lapse video microscopy of cellulose assemby by *Acetobacter xylinum*. J Cell Biol 109 (No 4, Pt 2) 90a.

Bureau TE, Brown Jr RM (1987) *In vitro* synthesis of cellulose II from a cytoplasmic membrane fraction of *Acetobacter xylinum*. Proc Natl Acad Sci USA 84:6985-6989.

Canale-Parola E, Borasky R, Wolfe RS (1961) Studies on *Sarcina ventriculi*. III. Localization of cellulose. J Bacteriol 81:311-318.

Deinema MH, Zevenhuizen LPTM (1971) Formation of cellulose fibrils by gram-negative bacteria and their role in bacterial flocculation. Archiv Mikrobiol 78:42-57.

Duchesne LC, Larson DW (1989) Cellulose and the evolution of plant life. BioScience 39(4):238-241.

Frey-Wyssling A (1976). In: Zimmerman W, Carlquist S, Ozenda P, Wulff HD (eds) The Plant Cell Wall. Gebruder Borntraeger, Berlin Stuttgart, p 277.

Gibbs SP (1981) The chloroplasts of some algal groups may have evolved from endosymbiotic eukaryotic algae. Ann New York Acad Sci 361:193-208.

Giddings TH, Brower DL, Staehelin LA (1980) Visualization of particle complexes in the plasma membrane of Micrasterias denticulata associated with the formation of cellulose fibrils. J Cell Biol 84:327-339.

Glaser L (1958) The synthesis of cellulose in cell-free extracts of Acetobacter xylinum. J Biol Chem 232:627-636.

Gunderson JH, Elwood H, Ingold A, Kindle K, Sogin ML (1987) Phylogenetic relationships between chlorophytes, chrysophytes, and oomycetes. Proc Natl Acad Sci USA 84:5823-5827.

Haigler CH, Brown Jr RM, Benziman M (1980) Calcofluor white ST alters cellulose synthesis in Acetobacter xylinum. Science 210:903-906.

Hall DA, Happey F, Lloyd PJ, Saxl H (1959) Oriented cellulose as a component of mammalian tissue. Proc R Soc London Ser B 151:497-516.

Hotchkiss Jr AT, Brown Jr RM (1987) The Association of rosette and globule terminal complexes with cellulose microfibril assembly in Nitella translucens var. axillaris (Charophyceae). J Phycol 23:229-237.

Hotchkiss A, and Brown Jr RM (1989) Evolution of the cellulosic cell wall in the charophyceae. In: Schuerch C (ed) Cellulose and Wood - Chemistry and Technology. John Wiley & Sons, New York, p 591-609.

Hotchkiss A, Gretz MR, Hicks KB, Brown Jr RM (1989) The composition and phylogenetic significance of the Mougeotia (Charophyceae) cell wall. J Phycol In press.

Lin FC, Brown Jr RM, Cooper JB, Delmer DP (1985) Synthesis of fibrils in vitro by a solubilized cellulose synthase from Acetobacter xylinum. Science 230:822-825.

Matthysee AG (1983) Role of bacterial cellulose fibrils in Agrobacterium tumefaciens infection. J Bacteriol 154(2) :906-915.

McLean B, Juniper BE (1986) The plasma membrane of young Chara internodal cells revealed by rapid freezing. Planta 169:153-161.

Mizuta S, Roberts EM, Brown Jr RM, (1989) A new cellulose synthesizing complex in Vaucheria hamata and its relation to microfibril assembly. In: Schuerch, C (ed) Cellulose and Wood - Chemistry and Technology. John Wiley & Sons, New York, p 659-676.

Moor H, Muhlethaler K (1963) Fine structure in frozen-etched yeast cells. J Cell Biol 17:609-628.

Mueller SC, Brown Jr RM, Scott TK (1976) Cellulosic microfibrils: nascent stages of synthesis in a higher plant cell. Science 194:949-951.

Mueller SC, Brown Jr RM (1980) Evidence for an Intramembranous component associated with a cellulose microfibril synthesizing complex in higher plants. J Cell Biol 84:315-326.

Pearasso R, Baroin A., Qu LH, Bachellerie JP, Adoutte A. (1989) Origin of the algae. Nature 339:142-144.

Pickett-Heaps JD, Marchant H (1972) The phylogeny of the green algae: a new proposal. Cytobios 6:255-264.

Preston RD (1964) Structural and mechanical aspects of plant cell walls with particular reference to synthesis and growth. In: Zimmermann, MH (ed) Formation of Wood in Forest Trees. Academic Press, New York, p 169-201.

Preston RD (1974) The physical biology of plant cell walls. Chapman and Hall, London.

Roberts EM, Saxena IM, Brown Jr RM (1989) Biosynthesis of Cellulose II in <u>Acetobacter xylinum</u>. In: Schuerch C (ed) Cell Wall and Wood - Chemistry and Technology. John Wiley and Sons, New York, p 689-704.

Roelofsen A (1958) Cell-wall structure as related to surface growth. Acta Bot Neerl 7:77-89.

Rolfe BG, Gresshoff PM (1988) Genetic analysis of legume nodule initiation. Ann Rev Plant Physiol and Plant Mol Biol 39:297-319.

Ross P, Aloni Y, Weinhouse C, Michaeli D, Weinberger-Ohana P, Meyer R, Benziman M (1985) An unusual oligonucleotide regulates cellulose synthesis in <u>Acetobacter xylinum</u>. FEBS Lett 186(2):191-196.

Saxena IM, Brown Jr RM (1989) Cellulose biosynthesis in <u>Acetobacter xylinum</u>: a genetic approach. In: Schuerch C (ed) Cellulose and Wood - Chemistry and Technology. John Wiley and Sons, New York, p 537-557.

Schramm M, Hestrin S (1954) Factors affecting production of cellulose at the air/liquid interface of a culture of <u>Acetobacter xylinum</u>. J Gen Microbiol 11:123-129.

Sisson W (1938) The existence of mercerized cellulose and its orientation in <u>Halicystis</u> as indicated by x-ray diffraction analysis. Science 87:350-351.

Stewart KD, Mattox KR (1978) Structural evolution in the flagellated cells of green algae and land plants. Biosystems 10:145-152.

Stewart KD, Mattox KR (1982) Phylogeny of phytoflagellates. In Rosowski JR, and Parker BC (ed) Selected Papers in Phycology II. Phycological Soc Amer, Lawrence, Kan p 626-640.

Wardrop AB (1970) The structure and formation of the test of Pyura stolonifera (Tunicata). Protoplasma 70:73-86.

Woese CR (1987) Bacterial evolution. Microbiol Reviews 51(2):221-271.

Yang D, Oyaizu Y, Oyaizu H, Olsen GJ, Woese CR (1985) Mitochondrial origins. Proc Natl Acad Sci USA 82:4443-4447.

Surface interactions in lichens

Rosmarie Honegger
University of Zürich
Institute of Plant Biology
Zollikerstr. 107
CH-8008 Zürich Switzerland

LICHEN PHOTOBIONTS AND THE LICHEN SYMBIOSIS

A considerable number of algal species inhabit a quite specialised ecological niche: the thallus of lichen-forming fungi. About 21% of fungi live as lichens in an ecologically obligate symbiosis with algal and/or cyanobacterial photobionts (Hawksworth 1988). More than 85% of lichen-forming fungi associate with unicellular or filamentous green algae, and about 10% with heterocyst-forming cyanobacteria, which provide photosynthates and, in the latter case, fixed nitrogen as well. About 3-4% of lichen mycobionts, the so-called cephalodiate species, associate with green and blue-green photobionts simultaneously (Tschermak-Woess 1988). In contrast to the genetically uniform thalli of multicellular algae, lichen thalli represent genetically heterogenous systems; the exact number of participants being unknown since several fungal and algal or cyanobacterial genotypes, originating from different sexual or asexual propagules, can and often do, participate in the formation of individual thalli (Jahns 1988). The species name given to lichens refers to the mycobiont (Art. 13.1d in Greuter 1988). Several investigators conclude from biogeographical studies that some ascomycetous lichen families might have arisen in palaeozoic times (summary: Hawksworth 1988). However, fossil records are largely missing because most lichen thalli contain no degradation-resistant biopolymers.

In less than 2% of all lichen species has the photobiont ever been determined at the species level (axenic culture under defined conditions is required), and often not even the generic affiliation is known (Tschermak-Woess 1988). About 100 species of lichen photobionts have so far been identified (Tschermak-Woess 1988). These fall into 7 orders of the Chlorophyta (Table 1), in one phaeophycean and one xanthophycean genus, and into 4 orders of cyanobacteria.

Table 1. **GENERA OF GREEN-ALGAL LICHEN PHOTOBIONTS**

abundance	genus	occurrence [1]	cell wall type
	CHAROPHYCEAE		
	Klebsormidiales		
***	Stichococcus NÄGELI	ph(as) + ns	n.d
	CHLOROPHYCEAE		
	Chlorococcales		
* ?	Asterochloris TSCHERMAK-WOESS	ph	n.d
**	Chlorella BEIJERINCK	ph(as) + ns	n.d [11]
*	Chlorosarcinopsis HERNDON [2]	ph(as) + ns	n.d
*** ?	Coccomyxa SCHMIDLE	ph(as) + ns	sporopollenin [9]
*** ?	Dictyochloropsis GEITLER	ph(as) + ns	intermediate [9,12]
** ?	Gloeocystis NÄGELI	ph + ns	n.d.
** ?	Elliptochloris TSCHERMAK-WOESS	ph	sporopollenin [9]
*** ?	Myrmecia PRINTZ	ph(as) + ns	intermediate [9,13]
** ?	Pseudochlorella LUND	ph + ns	intermediate [9]
** ?	Trochiscia KÜTZING [3]	ph(as) + ns	n.d
	ULVOPHYCEAE		
	Cladophorales		cellulose [10]
*	Cladophora KÜTZING	ph(as) + ns	n.d
	Ulotrichales		
** ?	Apatococcus BRAND [4] (syn. Protococcus C.AGARDH)	ph + ns	n.d
*	Blidingia KYLIN	ph(as) + ns	n.d
* ?	Coccobotrys CHODAT	ph	n.d
**	Pseudendoclonium WILLE [5] (syn. Dilabifilum TSCHERMAK-WOESS)	ph(as) + ns	n.d
	Prasiolales		xylo-mannan [10]
*	Prasiola AGARDH [6]	ph(as) + ns	n.d
	Pleurastrales		
**	Pleurastrum CHODAT (syn. Leptosira BORZI)	ph(as) + ns	n.d
✱✱	Trebouxia DE PUYMALY [7]	ph(as§)+ ns	cellulose or cellulose-like [9,14,15]
	Trentepohliales		
**	Cephaleuros KUNZE	ph(as) + ns	n.d
***	Phycopeltis MILLARDET	ph(as) + ns	n.d [16]
✱	Trentepohlia MARTIUS	ph(as) + ns	cellulose [9]
	INCERTAE SEDIS		
* ?	Nannochloris [8]	ph + ns	n.d

Sources: Tschermak-Woess (1988); Bourrelly (1972); various investigators in Irvine & John (eds., 1984); Sluiman (1989).

- * so far found as photobiont of a single lichen species
- ** so far found as photobiont(s) of less than 10 lichen spp.
- *** so far found as photobiont(s) of less than 100 lichen spp.
- ✱ very widespread in a large number of lichen spp.
- ✱✱ the most common group of lichen photobionts (in 50-70% of all lichens, as estimated by Ahmadjian 1982).
- ? taxa in which the main characteristics of modern systematics of green algae have not yet or incompletely been investigated.
- ph occurring as photobiont(s).
- (as) lichen photobionts some (or all) of which have been found in the aposymbiotic ("free-living") state in nature.
- ns non-symbiotic algae.
- n.d. cell wall structure and composition of the lichen photobionts not investigated.
- § Trebouxia spp. do occur in the aposymbiotic state, but they are rare (see Tschermak-Woess 1988)[but so are experts who are able to identify them!]

1 according to Tschermak-Woess (1988) and Bourrelly (1972)
2 higher taxonomic rank: Deason & Floyd (1987)
3 higher taxonomic rank: Komarek & Fott (1983)
4 Gärtner & Ingolić (1989).
5 John & Johnson (1989)
6 higher taxonomic rank: O'Kelly et al. (1989)
7 inclusive Pseudotrebouxia Archibald; see Gärtner (1985), Tschermak-Woess (1989). Fine Structure: Melkonian & Peveling (1988), Sluiman & Lokhorst (1988).
8 The photobiont of Normandina pulchella (a monotypic ascomycetous genus incertae sedis) was described as Nannochloris normandinae (Tschermak-Woess 1981) and integrated in the Chlorosarcinales (Tschermak- Woess 1981, 1988). TEM studies of the symbiotic state (Honegger, unpubl.) indicate that this alga is unlikely to be closely related to the non-symbiotic, chlorococcalean members of the genus Nannochloris NAUMANN as depicted e.g. by Menzel & Wild (1989).
9 cell wall types according to Brunner (1985) and Brunner & Honegger (1985) [see Table 3].
10 review: Percival & McDowell (1981).
11 at least 3 different cell wall types have been described in the genus Chlorella (sporopollenin-containing and others; Yamada & Sakaguchi 1982), but lichen photobionts have not yet been examined.
12 intermediate wall types reveal structural similarity with sporopollenin-containing walls, but contain no polymer which would resist non-oxidative degradation in their outermost wall layer.
13 reports on sporopollenin (Honegger & Brunner 1981) are based on a misdentification of the photobiont of Baeomyces rufus which was identified as Elliptochloris bilobata (Tschermak-Woess 1985).
14 König & Peveling (1984)
15 reports on sporopollenin (König & Peveling 1980, 1984) seem to be based on chemically inappropriate preparative procedures (Brunner 1985; Brunner & Honegger 1985).
16 reports on a sporopollenin-containing zone (Good & Chapman 1978) seem to be based on chemically inappropriate preparative procedures (Brunner 1985; Brunner & Honegger 1985).

➡ Note: The taxonomy of green algae is presently subjected to major changes. Only very few investigators have ever been examining lichen photobionts, and modern cell biological methods have seldom been applied in these studies. The present list is a preliminary attempt to integrate available data in modern taxonomic concepts.

Considering the taxonomic diversity of the organisms involved, the polyphyletic origins and the ancient phylogenetic age of some of these symbiotic systems, it is not surprising that many different types of mycobiont- photobiont interactions have developed amongst lichens. In less than 1% of all lichen spp. has the symbiotic relationship between the partners been studied with cell biological and/or physiological methods, and most of these investigations focus on few highly advanced taxa whose characteristics are:

- three-dimensional thalli built up by the mycobiont with distinct, often species-specific external morphology (squamulose, foliose or fruticose forms). The mycobiont as exhabitant is the quantitatively predominant partner and makes up more than 90% of the thalline biomass.
- internal stratification; the photobiont cell-population being kept in a particular layer in the gas-filled, usually internal, zone of the thallus and controlled by the mycobiont. The entire photobiont cell-population depends on water and nutrient supply by the mycobiont (Fig. 7).
- positioning of the photobiont cells within the thallus either by specialized hyphae, e.g. the intragelatinous protrusions in cyanobionts in the Peltigeraceae/Peltigerales, or by very characteristic haustorial complexes in lecanoralean species with trebouxioid photobionts (summary: Honegger 1988). These mycobiont-photobiont interfaces are the site of bidirectional fluxes. Water, minerals and fungal metabolites coming from the fungal side, and photosynthates from the photobiont, e.g. an acyclic polyol from green algal photobionts, or a glucan from cyanobionts (summary: Smith & Douglas 1987).
- moderate to high specificity towards the green algal photobiont (Table 2).
- reduction of the generative reproductive cycle in a relatively high percentage of mycobiont spp. which disperse very efficiently by means of asexually-produced symbiotic diaspores (soredia, isidia etc). This reproductive strategy allowed the successful colonisation of terrestrial ecosystems in which compatible photobionts do not occur abundantly in the aposymbiotic state.

Highly advanced lichens represent mutualistic symbioses, the biological fitness of both partners being increased in the symbiotic state (as concluded from the ecological breadth and biomass of both partners). They are the product of a long cohabitation of the symbionts which has led to a high degree of specialisation in the fungal partner while the the photobiont has remained simple. Law and Lewis (1983) postulated that there was a selection against sexual reproduction, leading to a low rate of genetic interchange in inhabitants of mutualistic symbioses with a substantial history of co-evolution. Exhabitants of such symbioses would retain the sexual reproductive cycle. At first sight the lichen symbiosis appears to fit remarkably well into this scheme. However, non-symbiotic relatives of many of the green-algal lichen photobionts do not reproduce sexually either. Moreover, highly advanced lichen mycobionts show a remarkable trend to reduce the sexual reproduction in favour of asexual propagation of the symbiotic state.

Besides these highly advanced taxa there are large numbers of primitive lichens in most of which the symbiotic relationship is not understood. Many of these primitive lichen symbioses are probably more or less mild forms of fungal parasitism on algal or cyanobacterial hosts. Characteristics of primitive lichens are:

- no 3-dimensional thalli with incorporated photobiont cells, but ensheathment of compatible algal or cyanobacterial cells or colonies on or within the substrate; the result being a crust-like thallus with no distinct internal stratification. The mycobiont may or may not be the quantitatively predominant partner.
- intracellular fungal haustoria are commonly formed in cyanobacterial and in green algal photobionts with cellulosic or intermediate cell walls (Tab.1), but usually not in walls which contain sporopollenin (Tab. 3);
- very low to moderate specificity towards the photobiont (Tab. 2) (Tschermak-Woess 1976, 1988).
- the generative reproductive cycle is the most common mode of dispersal; symbiotic diaspores are less frequently formed than in highly advanced taxa.

These primitive forms can be regarded as the "testing ground" of many taxonomically unrelated fungi which have discovered independently that forming a biotrophic association with algal or cyanobacterial photobionts is a convenient mode of nutrition.

RECOGNITION, SPECIFICITY

Very little experimental work has been carried out on specificity and recognition in the lichen symbiosis. Resynthesis experiments with axenically cultured aposymbiotic partners are not performed easily and have so far only seldom resulted in symbiotic phenotypes whose morphology and anatomy was identical with naturally grown thalli; see the pioneering work of V. Ahmadjian (e.g. Ahmadjian & Jacobs 1987). Various degrees of specificity ranging from very low to high have so far been observed in lichens (Table 2); moderate specificity probably being the most common situation. However, the estimated 50-70% of all lichen-forming fungi which associate with trebouxioid phycobionts may be little- to moderately-specific, but they are highly selective as regards to their photobionts since Trebouxia spp. are not commonly found in all habitats which can support lichen growth (Tschermak-Woess 1988). Most of the

Table 2. The degree of taxonomic difference between acceptable lichen photobionts as a measure for the degree of SPECIFICITY IN A RANGE OF LICHEN-FORMING ASCOMYCETES*

VERY LOW SPECIFICITY (acceptable photobionts found within the same phylum)

e.g. **Chaenotheca carthusiae** (Caliciales) (Tibell, 1982)
crustose species with non-stratified thallus

photobionts:		
	Trebouxia sp. (Pleurastrales/Ulvophyceae)	N-Hemisphere
	Stichococcus sp. (Klebsormidiales/Charophyceae)	Costa Rica

MODERATE SPECIFICITY (acceptable photobionts found within the same genus)

→ PROBABLY THE MOST COMMON SITUATION IN FOLIOSE AND FRUTICOSE TAXA

e.g. **Xanthoria parietina** (Teloschistales) (summary: Gärtner, 1985)
foliose species with internally stratified thallus

photobionts:		
	Trebouxia aggregata	Cechoslovakia, Belgium
	T. decolorans (incl. T. albulescens)	Italy
	T. italiana	Italy

HIGH SPECIFICITY (acceptable photobionts found within the same species)

e.g. **Pseudocyphellaria aurata** (Peltigerales) (Tschermak-Woess, 1984)
foliose species with internally stratified thallus

photobionts:		
	Dictyochloropsis symbiontica var. symbiontica	Tenerife, 600 m alt.
		New Zealand, 240 m
	var. pauciautosporica	Tenerife, 700-800 m

VERY HIGH SPECIFICITY (acceptable photobiont found within the same subspecific taxon)

NOT (YET) ENCOUNTERED IN THE LICHEN SYMBIOSIS

* data based on investigations of naturally grown lichen thalli; experimental studies (resynthesis experiments with a range of photobionts) are largely missing. Further examples: see Tschermak-Woess (1988).

Table 3. **CELL WALL PROPERTIES** in representatives of 3 genera of green-algal lichen photobionts

TREBOUXIA spp.	**COCCOMYXA and ELLIPTOCHLORIS spp.**
Cell wall structure and composition [1,3-7]	**Cell wall structure and composition** [1,2,5,6,8]
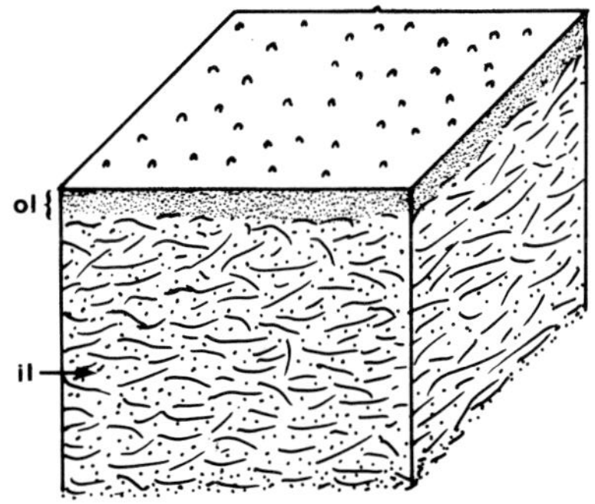	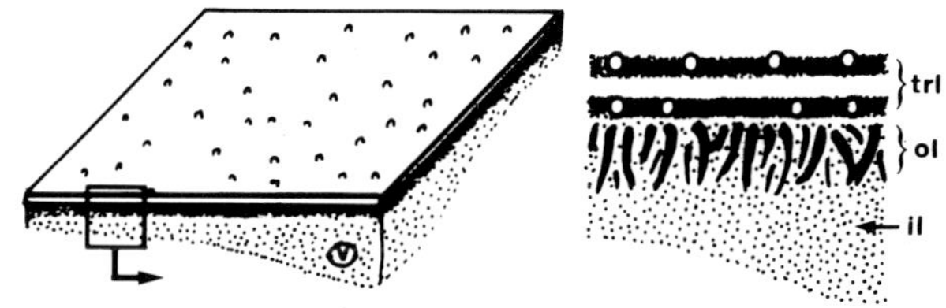
ol amorphous outer wall layer (30 ± 15 nm) containing proteinaceous particles embedded in an amorphous matrix of acidic polysaccharides il inner wall layer (1000 ± 150 nm) built up mainly by an irregular meshwork of cellulosic or cellulose-like fibres	trl trilaminar outermost wall layer (15 ± 5 nm) containing sporopollenin, probably in its central part, and (probably proteinaceous) particles on its outer and inner surfaces. Sporopollenin content: 6-11% of wall dry weight ol outer wall layer (30 ± 10 nm) with short, cellulose-like fibres embedded in an amorphous matrix. This layer appears electron-dense in conventionally fixed specimens il amorphous inner layer (20 - several hundreds of nm), sometimes with golgi-derived vesicles (v); its size and shape may partly be a result of swelling/ shrinkage artifacts in the cell during conventional preparative procedures
Special cell wall properties [1,5]	**Special cell wall properties** [1,5]
Disruption of cell walls by pressure shock treatment (French Press) possible	Disruption of cell walls by ballistic treatment only (e.g. with a Braun MSK homogenizer)
Isolated cell walls are like soft bags; flat spreading on mica surfaces and subsequent replication possible	Isolated walls are like brittle shells with convolute margins. Spreading and replication impossible
Cell walls are diffusible for relatively large molecules (e.g. aniline blue, molecular weight 738)	Cell wall not diffusible for larger molecules (e.g. aniline blue, molecular weight 738)
Mycobiont-photobiont relationships [5-7,10,11]	**Mycobiont-photobiont relationships** [5-7,10,11]
Intracellular and intraparietal fungal haustoria are common and widespread	Wall-to-wall apposition is the most common mode of interaction; no haustoria are formed, not even in crustose forms or in goniocysts of basidiolichens. Exception: goniocysts of *Micarea prasina*.
Mother cell walls are enzymatically degraded	Trilaminar sheath of mother cell wall persists, the rest is degraded.
Photobionts of many of the anciently lichenized taxa, of many primitive and of the highest advanced lichen mycobionts (e.g. the Parmeliaceae)	Photobionts of many Peltigeraceae (which are all associating simultaneously with cyanobacteria), and of some small groups of unrelated lichen-forming asco-and basidiomycetes

1: Brunner 1985; 2: Brunner & Honegger 1985; 3: Bubrick & Galun 1980; 4: König & Peveling 1984; 5: Honegger 1984; 6: Honegger 1985; 7: Honegger 1986a; 8: Honegger & Brunner 1981; 9: Oberwinkler 1984; 10: Plessl 1963; 11: Tschermak 1941.

very common and widespread terrestrial unicellular green algae are not compatible photobionts, and the most frequently observed green-algal lichen photobionts are rarely found in the aposymbiotic state (Galun and Bubrick 1984). Apparently, many of the lichen mycobionts which disperse exclusively by ascospores, and thus have to re-establish the symbiotic relationship at each reproductive cycle, can form crust-like, non-stratified pre-thalli (also termed "soredial stage", "basal tissue" etc.) with a range of incompatible algal cells until compatible photobionts become available. Only compatible photobionts trigger, in an unknown manner, the expression of the symbiotic phenotype ("morphotype") of the fungal genotype. Aposymbiotic fungal phenotypes, as observed in axenically cultured lichen mycobionts, are completely different morphologically from symbiotic ones. This very interesting survival strategy was observed in axenic resyntheses of mycobionts with a range of compatible and incompatible photobionts (Ahmadjian & Jacobs 1981) and in naturally grown thalli (Ott 1987).

PHOTOBIONT CELL WALL PROPERTIES AND MYCOBIONT-PHOTOBIONT RELATIONSHIPS

Recognition of compatible photobionts and the establishment of a stable symbiosis is a multistep process. Like in other symbiotic systems the structure and composition of the photobiont cell wall play a key role in establishing a symbiotic relationship (Tab. 3). It seems particularly interesting that the very thin (15±5 nm), outermost trilaminar wall layer of Coccomyxa and Elliptochloris spp. largely protects the cells from fungal invasion due to its sporopollenin content (Honegger & Brunner 1981; Honegger 1984; Brunner 1985; Brunner & Honegger 1985). This extremely chemically resistant wall component belongs to a class of biopolymers with an interesting distribution among microorganisms (summary: Honegger & Brunner 1985). Moreover, it is a major component of microspore walls in plants. Sporopollenin preparations of aposymbiotically cultured lichen photobionts were obtained by incubating photobiont cell walls which had been exhaustively extracted with a range of organic solvents, then saponified under alcaline conditions, in 85% phosphoric acid at 50-55°C for 13-30 days (Brunner & Honegger 1985). The thin, trilaminar sheath of Coccomyxa and Elliptochloris spp. resisted these procedures, whereas walls of Dictyochloropsis, Myrmecia, Pseudochlorella, Trebouxia and Trentepohlia spp contained no degradation-resistant material. However, all of these photobiont cell walls yielded small amounts of resistant material, when incompletely extracted samples were subjected to acetolysis, a chemically less than optimal, but widely used preparative procedure (Brunner 1985; Brunner & Honegger 1985). Acetolysis not only changes the physico-chemical properties of natural sporopollenin (Berkaloff et al. 1983), but is likely to polymerize wall-bound carotenoids of incompletely extracted samples to a resistant, sporopollenin-like material whose IR-spectra are almost identical with those of acetolysed Lycopodium sporopollenin. Thus, various reports on the occurrence of a sporopollenin-like biopolymer in algal and other cell walls are likely to be based on method-dependent artifacts (summary: Brunner & Honegger 1985).

The infra-red spectra of Coccomyxa and Elliptochloris sporopollenins were almost identical with those of the hydrocarbon-rich, chlorococcalean Botryococcus braunii (Berkaloff et al. 1983), but differed from Lycopodium spore wall sporopollenin, which was used as reference material by most investgators (Brunner 1985; Brunner & Honegger 1985). ^{13}C NMR investigations of Guilford et al. (1989) indicate that algal wall sporopollenins are derived from fatty acids rather than from carotenoids and/or carotenoid esters, as previously assumed (e.g. Burczyk 1987). Lycopodium sporopollenin, on the other hand, seems to be derived from carotenoids (Guilford et al. 1989).

Sporopollenin-containing cell walls have been described in a wide range of non-symbiotic and symbiotic unicellular green algae, the latter being endosymbionts of numerous protozoan and invertebrate hosts (e.g. Jolley & Smith 1980; Wolf & Cox 1981; Berkaloff et al. 1983; Burczyk 1987). A striking structural similarity of the cell walls of Dictyochloropsis, Myrmecia and Pseudochlorella spp. with those of Coccomyxa and Elliptochloris spp. was noted. However, only members of the latter two genera contained sporopollenin, whereas the trilaminar sheath of the others was devoid of a degradation-resistant biopolymer (Brunner, 1985; Brunner & Honegger 1985). Thus, the presence of a trilaminar outermost wall layer is not conclusive evidence for the occurrence of sporopollenin in algal cell walls.

A different cell wall structure and composition has been observed in lichen photobionts of the genera Trebouxia sensu lato and Trentepohlia (see Tab. 3). Cellulose or cellulose-like components have been found in all lichen photobionts so far examined (König & Peveling 1984; Honegger 1984; Brunner 1985). The residues of extracted cell walls yielded, on hydrolysis, glucose in Trentepohlia spp., or glucose and small amounts of mannose in representatives of the genera Coccomyxa, Elliptochloris, Myrmecia and Trebouxia (Brunner 1985). Glucose was found in the hydrolysates of 4, and a mixture of glucose and mannose in 2 out of 6 species of Trebouxia by König and Peveling (1984). No X-ray crystallographic data are available. Residues (after sequential extraction) which give predominantly glucose, but also appreciable amounts of mannose on hydrolysis, have been found in a wide range of green algae (Percival & McDowell 1981; Brunner 1985).

Surface interactions in lichens include an additional, quite fascinating element. In advanced lichen taxa with an internally stratified thallus and a gas-filled medullary layer the green algal or cyanobacterial cells are contacted by aerial hyphae of the mycobiont. The aerial hyphae at the mycobiont-photobiont interface and the medullary layer of lichens have, like aerial hyphae of non-lichenized fungi (e.g. hyphophores, conidia etc.), a very thin, more or less hydrophobic outermost cell wall layer which seems to function as a cuticle equivalent. These hydrophobic cell wall surface layers channel solutes which move passively within the apoplastic space of the cell wall, and reduce their loss. They show either a distinct rodlet pattern (Fig. 10), or an irregularly tessellated structure (Figs. 2, 6), probably depending on their composi-tion. Lipids, proteins, carbohydrates and phenolics have been identified by various investigators with chemical and histochemical techniques (summary: Honegger 1988). Such water-repellent fungal cell wall surface layers can spread easily over suitable substrates (Beever, pers. comm.).

In lecanoralean lichen mycobionts with trebouxioid photobionts the rodlet layer in the tip area of the growing fungal hypha spreads over the wall surface of the juvenile algal cell after the initial contact (Fig. 6). Subsequently the rodlet pattern of the fungal and algal wall surface becomes obscured by an amorphous material which contains lipids and phenolic secondary metabolites. This amorphous material increases the hydrophobicity of the outermost wall layer and provides it with an irregularly tessellated pattern (Figs. 2, 6; Honegger 1984, 1985, 1986b, 1988). Thus, each _Trebouxia_ cell is contacted not only by a haustorial complex, but is also sealed with mycobiont-derived water-repellent material. Sealing of the photobiont wall surface with hydrophobic material canalises the flux of solutes during the regular wetting and drying cycles of these poikilohydrous systems and facilitates the bi-directional flux at the interface during water saturation. Thus, the photobiont cell wall surface remains uncovered by free water even at water saturation, and this ensures unhindered gas exchange for photosynthesis. This situation was postulated on the basis of ultrastructural and histochemical studies (Honegger 1985, 1986b, 1988) and confirmed impressively by the first low temperature scanning electron microscopy (LTSEM) micrographs of wet and dry _Parmelia sulcata_ by A. Beckett and co-workers (Brown et al. 1987).

A different situation was observed at the interface of peltigeracean myco-

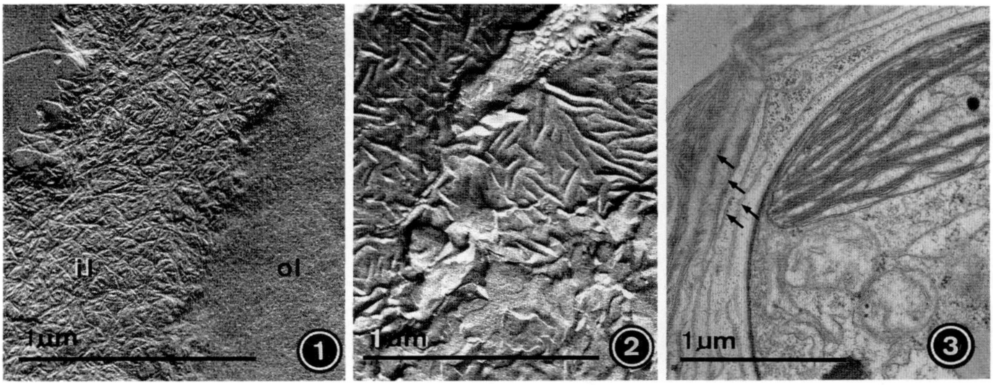

Figs. 1-3. Cell wall structure in photobionts of the genera Trebouxia and Coccomyxa. Fig. 1. TEM micrograph of a replica of an isolated cell wall of a cultured Trebouxia sp., showing an amorphous outer layer (ol), and an irregular meshwork of cellulose-like fibres in the inner layer (il). Fig. 2. TEM micrograph of a freeze-fractured specimen showing the cell wall surface the same Trebouxia sp. in the symbiotic state in the thallus of the lecanoralean lichen Cladonia macrophylla (Schaerer)Stenh.. The algal cell wall is covered by a mycobiont-derived, hydrophobic wall surface layer which shows an irregularly tessellated pattern (see Fig. 6). Fig. 3. TEM micrograph of an ultrathin section of the Coccomyxa photobiont of Icmadophila ericetorum (L.) Zahlbr. (Helotiales). Arrows point to sheaths of non-degradable, sporopollenin-containing mother cell walls.

Figs. 4-10. Surface interactions in lichens: mycobiont-derived cell wall surface layers on green-algal photobiont cell walls. Figs. 4-6. Cladoniaceae (Lecanorales) with Trebouxia photobiont. Fig. 4. SEM micrograph of Cladonia arbuscula (Wallr.)Rabenh., showing the mycobiont-photobiont interface. Fig. 5. TEM micrograph of the mycobiont-photobiont interface in Cladonia caespiticia (Pers.)Flörke, showing a median (left) and a tangential (right) section of an intraparietal haustorium (iph) of the mycobiont (MY) in the cell wall of the Trebouxia photobiont (PH). sl: electron-dense, mycobiont-derived cell wall surface layer. Fig. 6: Freeze-fracture preparation of Cladonia macrophylla, showing the interface between growing mycobiont (MY) hyphae (the top one was torn away during freeze-cleaving) and the wall surface of a juvenile photobiont (PH) cell. The smooth algal wall surface (see Fig. 1) is ensheathed by a thin, mycobiont-derived cell wall surface layer with a distinct rodlet pattern (r(MY)) which becomes soon obscured by overlaying, amorphous material, giving the wall surface an irregularly tessellated pattern (it). Figs. 7-10. Mycobiont-photobiont interface in Peltigeraceae (Peltigerales) with Coccomyxa photobiont. Figs. 7-8. SEM micrographs of the mycobiont-photobiont interface in Peltigera aphthosa (L.)Willd.; uc: upper cortical layer; PH: photobiont cell population; m: gas-filled medullary layer. Figs. 9-10. Freeze-fracture preparations of Solorina saccata (L.)Ach.. Fig. 9. Fracture following the sporopollenin-containing, outermost cell wall layer of 2 young autospores (PH1 and PH2), and the wall surface of the partly fractured, sporopollenin-containing mother cell wall (mcw) with adhering, mycobiont-derived cell wall surface layer with distinct rodlet pattern (r(MY)). Fig. 10. Fracture following the cleaved mycobiont cell wall (cw(MY)), and the mycobiont wall surface layer with distinct rodlet (r(MY)) pattern which adheres to the photobiont (PH) wall surface.

(Preparative procedures: see Honegger 1986b).

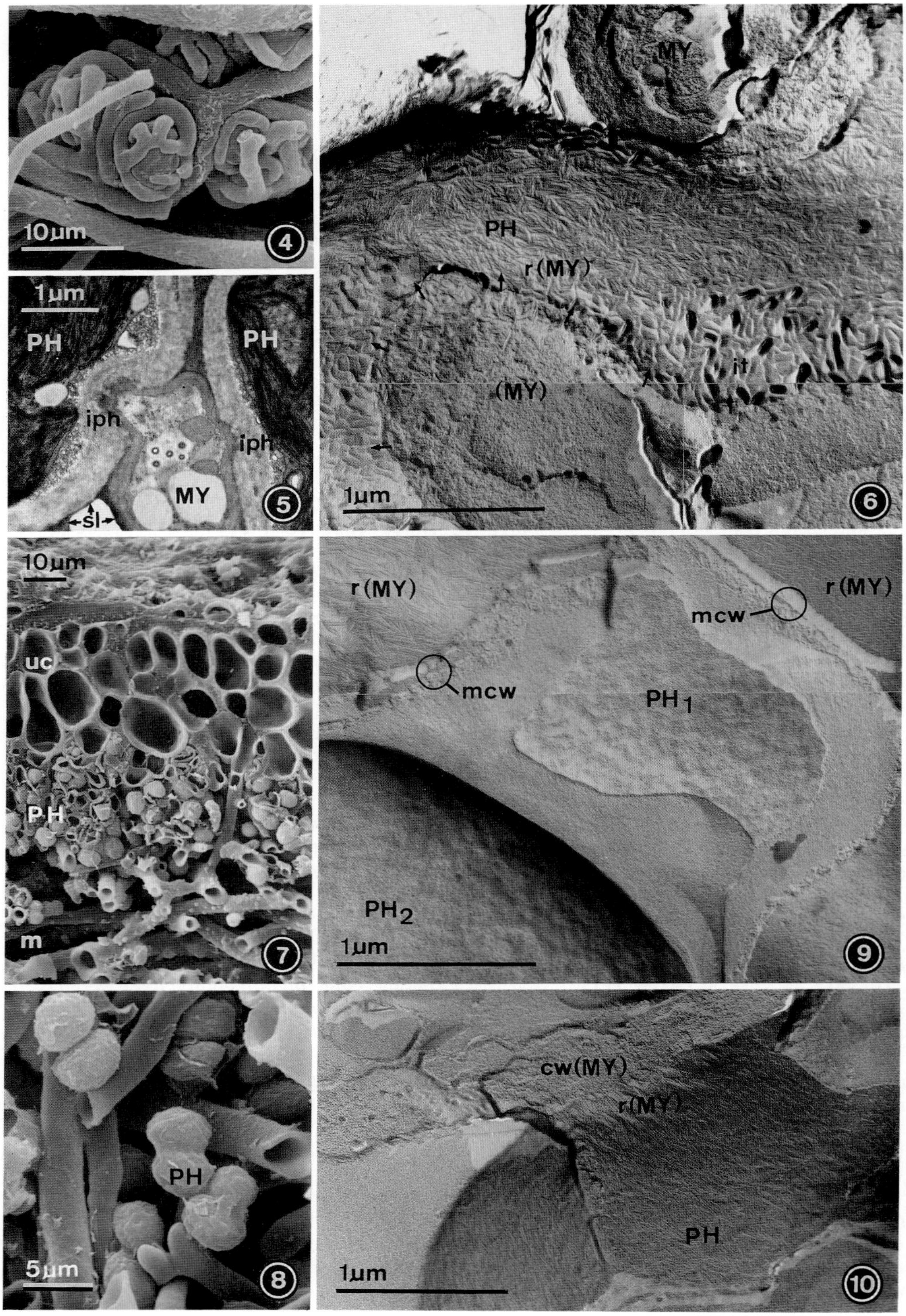
10µm
4
1µm
PH
PH
iph
iph
MY
sl
5
MY
PH
r(MY)
it
(MY)
1µm
6
10µm
uc
PH
m
7
r(MY)
r(MY)
mcw
mcw
PH1
PH2
1µm
9
PH
5µm
8
cw(MY)
r(MY)
PH
1µm
10

bionts with Coccomyxa photobionts (Figs. 9-10). In the Peltigeraceae the wall surface layer of the aerial hyphae of the medullary and algal layers reveal a distinct rodlet pattern (Honegger 1982, 1985). This rodlet layer neither spreads over the Coccomyxa cell wall surface, nor does it ensheath the photobiont cells, but it adheres tightly to their outermost, sporopollenin-containing wall layer at the contact site (Figs. 9-10). The rodlet layer of Peltigera and Solorina spp. is less hydrophobic than the irregularly tessellated wall surface layer of members of the Lecanorales. Histochemical tests suggest a higher protein content and distinctly less lipid and phenolics than in the Lecanorales (unpubl. results). The photobiont cell population of Peltigera thalli receives water, nutrients and probably also fungal metabolites via the mycobiont (Figs. 7-8); the possible modes and routes of translocation are not yet understood.

PROSPECTS

The symbiosis of advanced lichens differs considerably in many respects from protozoan and invertebrate symbioses with unicellular algae as C- autotrophic partners, but does share some important features with them. In all these symbiotic systems the algal cells are inhabitants, lichen photobionts being extracellular endosymbionts of multicellular fungal structures, whereas algal symbionts of protozoans and invertebrates are intracellularly located. In all of these symbiotic systems the algal cell population is controlled and regulated by the host. Neither killing and digesting, nor expulsion, but shortage of nutrients and the quantitatively significant export of photosynthates to the host, are likely to be the main control mechanisms which keep the photobiont cell population of all these symbioses stable over long periods of time (Smith & Douglas 1987; Hill 1985, 1989; Honegger 1987). In lichens, the above regulation is possibly combined with a mycobiont-derived inhibitory/ stimulatory principle (Honegger 1987).

Although of limited economic interest, the ecologically important symbioses of lichen-forming fungi and their algal and/or cyanobacterial photobionts can be regarded as interesting model systems for studying different aspects of mycobiont-photobiont relationships, mutualism and evolution. Phycologists are facing a whole series of interesting questions about lichen algae, their metabolism in the symbiotic and aposymbiotic state, their fine structure in general and the structure and composition of their walls in particular, and their position in modern taxonomic concepts.

ACKNOWLEDGEMENTS

My sincere thanks are due to Drs H.R. Preisig and H. Sluiman for invaluable help in the preparation of Table 1; to Mrs. Verena Kutasi for doing all the darkroom work; and to Prof. A. Slusarenko for reading this manuscript and improving the English.

REFERENCES

Ahmadjian V (1982) Algal/fungal symbioses. In: F. Round and D.J. Chapman (eds): Progress in Phycological Research. Elsevier Biomedical Press Amsterdam, pp. 179-233

Ahmadjian V, Jacobs JB (1981) Relationship between fungus and alga in the lichen *Cladonia cristatella* Tuck. Nature 389: 169-172

Ahmadjian V, Jacobs JB (1987) Studies on the development of synthetic lichens. Bibl Lichenol 25: 47-58

Berkaloff C, Casadevall E, Largeau C, Metzger P, Peracca S, Virlet J (1983) The resistant polymer of the walls of the hydrocarbon-rich alga *Botryococcus braunii*. Phytochemistry 22: 389-397.

Bourrelly P. (1972) Les algues de l' eau douce. Tome I: Les algues vertes. N. Boubée & Cie Paris

Brown DH, Rapsch S, Beckett A, Ascaso C (1987) The effect of desiccation on cell shape in the lichen *Parmelia sulcata* Taylor. New Phytol 105: 295-299

Brunner U (1985) Ultrastrukturelle und chemische Untersuchungen an Flechtenphycobionten aus 7 Gattungen der Chlorophyceae (Chlorophytina), unter besonderer Berücksichtigung sporopollenin-ähnlicher Biopolymere. Juris Druck & Verlag Zürich

Brunner U, Honegger R (1985) Chemical and ultrastructural studies on the distribution of sporopollenin-like biopolymers in 6 genera of lichen phycobionts. Can J Bot 63: 2221-2230

Bubrick P, Galun M (1980) Proteins from the lichen *Xanthoria parietina* which bind to phycobiont cell walls. Correlation between binding patterns and cell wall cytochemistry. Protoplasma 104: 167-173

Burczyk J (1987) Biogenetic relationships between ketocarotenoids and sporopollenin in green algae. Phytochemistry 26: 113-120

Deason TR, Floyd GF (1987) Comparative ultrastructure of three species of *Chlorosarcina* (Chlorosarcinaceae, Chlorophyta). J Phycol 23: 187-195

Galun M, Bubrick P (1984) Physiological interactions between the partners of the lichen symbiosis. In: HF Linskens, J Heslop-Harrison (eds): Cellular Interactions. Encyclopedia of Plant Physiology. Springer Berlin, pp. 362-401.

Gärtner G (1985) Die Gattung Trebouxia Puymaly (Chlorellales, Chlorophyceae). Arch Hydrobiol Suppl 71: 495-548

Gärtner G, Ingolic E (1989) Ein Beitrag zur Kenntnis von Apatococcus lobatus (Chlorophyta, Chaetophorales, Leptosiroideae). Pl Syst Evol 164: 133-143

Good BH, Chapman RL (1978) The ultrastructure of Phycopeltis (Chroolepidaceae: Chlorophyta). 1. Sporopollenin in the cell walls. Am J Bot 65: 27-33

Greuter W (ed) (1988) International Code of Botanical Nomenclature. Vol 118. Koeltz Königstein

Guilford WJ, Schneider DM, Labovitz J, Opella SJ (1988) High resolution solid state 13C NMR Spectroscopy of Sporopollenins from different plant taxa. Plant Physiol 86: 134-136

Hawksworth DL (1988) The fungal partner. In: M Galun (ed): Handbook of Lichenology. CRC Press Boca Raton, pp. 35-38

Hill DJ (1985) Changes in photobiont dimensions and numbers during co-development of lichen symbionts. In: DH Brown (ed): Lichen Physiology and Cell Biology. Plenum New York, pp. 303-317

Hill DJ (1989) The control of the cell cycle in microbial symbionts. New Phytol 112: 175-184

Honegger R (1982) Cytological aspects of the triple symbiosis in Peltigera aphthosa. J Hattori Bot Lab 52: 379-391

Honegger R (1984) Cytological aspects of the mycobiont-phycobiont relationship in lichens. Haustorial types, phycobiont cell wall types, and the ultrastructure of the cell wall surface layers in some cultured and symbiotic myco- and phycobionts. Lichenologist 16: 111-127

Honegger R (1985) Fine structure of different types of symbiotic relationships in lichens. In: DH Brown (ed): Lichen Physiology and Cell Biology. Plenum Press New York, pp. 287-302

Honegger R (1986a) Ultrastructural studies in lichens. I. Haustorial types and their frequencies in a range of lichens with trebouxioid phycobionts. New Phytol 103: 785-795

Honegger R (1986b) Ultrastructural studies in lichens. II. Mycobiont and photobiont cell wall surface layers and adhering crystalline lichen products in four Parmeliaceae. New Phytol 103: 797-808

Honegger R (1987) Questions about pattern formation in the algal layer of lichens with stratified (heteromerous) thalli. Bibl Lichenol 25: 59-71

Honegger R (1988) The functional morphology of cell-to-cell interactions in lichens. In: S Scannerini, DC Smith, P Bonfante-Fasolo, V Gianinazzi-Pearson (eds): Cell to cell signals in plant, animal and microbial symbiosis. Springer/NATO ASI series Berlin, Series H, vol 17, pp. 39-53

Honegger R, Brunner U (1981) Sporopollenin in the cell wall of Coccomyxa and Myrmecia phycobionts of various lichens: an ultrastructural and chemical investigation. Can J Bot 59: 2713-2734

Irvine DEG, John DM (eds) (1984) Systematics of the green algae. Academic Press London, pp. 1-449

Jahns HM (1988) The lichen thallus. In: M Galun (ed): Handbook of Lichenology. CRC Press Boca Raton, pp. 95-143

John DM, Johnson LR (1989) A cultural assessment of the freshwater species of Pseudendoclonium WILLE (Ulotrichales, Ulvophyceae, Chlorophyta). Arch Hydrobiol Suppl 82: 79-112

Jolley E, Smith DC (1980) The green hydra symbiosis. II. The biology of the establishment of the association. Proc R Soc London Ser. B 207: 311-333

Komarek J, Fott B (1983) Die Binnengewässer Bd. XVI. Das Phytoplankton des Süsswassers, 7. Teil, 1. Hälfte. Chlorophyceae (Grünalgen), Ordnung Chlorococcales. Stuttgart: Schweizerbart'sche Verlagsbuchhandlung

König J, Peveling E (1980) Vorkommen von Sporopollenin in der Zellwand des Phycobionten Trebouxia. Z Pflanzenphysiol 98: 459-464

König J, Peveling E (1984) Cell walls of the phycobionts Trebouxia and Pseudotrebouxia: constituents and their location. Lichenologist 16: 129-144

Law R, Lewis D (1983) Biotic environments and the maintenance of sex - some evidence from mutualistic symbioses. Biol J Linn Soc 20: 249-271

Melkonian M, Peveling E (1988) Zoospore ultrastructure in species of Trebouxia and Pseudotrebouxia (Chlorophyta). Pl Syst Evol 158: 183-210

Menzel K, Wild A (1989) A comparative ultrastructural investigation of some Nannochloris species (Chlorococcales) with particular reference to the systematic position of Nanochlorum eucaryotum. Bot Acta 102: 152-158

O'Kelly CJ, Garbary DJ, Floyd GL (1989) Flagellar apparatus of male gametes and other aspects of gamete and zygote ultrastructure in Prasiola and Rosenvingiella (Chlorophyta, Prasiolales) from British Columbia. Can J Bot 67: 505-514

Oberwinkler F (1984) Fungus-alga interactions in basidiolichens. Nova Hedwigia Beih 79: 739-774

Ott S (1987) Reproductive strategies in lichens. Bibl Lichenol 25: 81-93

Percival E, McDowell RH (1981) Algal walls - composition and biosynthesis. In: W Tanner, FA Loewus (eds): Plant Carbohydrates II. Extracellular Carbohydrates. Springer Berlin, pp. 277-316

Plessl A (1963) Ueber die Beziehungen von Pilz und Alge im Flechtenthallus. Oesterr Bot Z 110: 194-269

Sluiman HJ (1989) The green algal class Ulvophyceae. An ultrastructural survey and classification. Crypt Bot 1: 83-94

Sluiman HJ, Lokhorst GM (1988) The ultrastructure of cellular division (autosporogenesis) in the coccoid green alga, Trebouxia aggregata, revealed by rapid freeze fixation and freeze substitution. Protoplasma 144: 149-159

Smith DC, Douglas AE (1987) The Biology of Symbiosis. Edward Arnold London

Tibell L (1982) Caliciales of Costa Rica. Lichenologist 14: 219-254

Tschermak E (1941) Untersuchungen über die Beziehungen von Pilz und Alge im Flechtenthallus. Oesterr Bot Z 90: 233-307

Tschermak-Woess E (1976) Algal taxonomy of lichens: the phycobiont of Verrucaria adriatica. In: DH Brown, DL Hawksworth, RH Bailey (eds): Lichenology: Progress and Problems. Academic Press London, pp. 79-88

Tschermak-Woess E (1981) Zur Kenntnis der Phycobionten von Lobaria linita und Normandina pulchella. Nova Hedwigia 35: 63-73

Tschermak-Woess E (1984) Ueber die weite Verbreitung lichenisierter Sippen von Dictyochloropsis und die systematische Stellung von Myrmecia reticulata (Chlorophyta). Pl Syst Evol 147: 299-322

Tschermak-Woess E (1985) Elliptochloris bilobata, kein ganz seltener Phycobiont. Herzogia 7: 105-116

Tschermak-Woess E (1988) The algal partner. In: M Galun (ed): Handbook of Lichenology. CRC Press Boca Raton, pp. 39-92

Tschermak-Woess E (1989) Developmental studies in trebouxioid algae and taxonomic consequences. Pl Syst Evol 164: 161-195

Wolf FR, Cox ER (1981) Ultrastructure of active and resting colonies of Botryococcus braunii (Chlorophyceae). J Phycol 17: 395-405

Yamada T, Sakaguchi K (1982) Comparative studies on Chlorella cell walls: induction of protoplast formation. Arch Microbiol 132: 10-13

Participation of algal cell wall surface structures in the formation of the host-symbiont-interface of endocytobiotic systems

W. Reisser
Pflanzenphysiologisches Institut und Botanischer Garten
University of Göttingen
Untere Karspüle 2
D - 3400 Göttingen
West Germany

1. Characteristics Of The Host-Symbiont-Interface In Endocytobiotic Systems

Cell wall surfaces play an important role in a broad range of interactions of algae with their biotic and abiotic environment. Cell surface components are responsible for adhesion of algae to special substrates, can protect them against desiccation and show antibiotic properties. They participate in sexual processes among algae as well as in colony formation and are of key importance for all kind of epiphytic, parasitic, and symbiotic relationships of algae.

In symbiotic relationships, the algal cell wall surface is part of the host-symbiont-interface which may be defined by both structure and function: It is the area of physical contact between partners through which they exchange a broad range of signals. Accordingly, symbiotic systems formed by algae can be classified by the structure of the host-symbiont-interface (Tab. 1). In endocytobiotic systems with eukaryotic algae the interface belongs to the cell wall-membrane type. Well studied systems of that type are the associations of ciliate and *Hydra* species with chlorophycean (Reisser 1986, Mc Neil and McAuley 1984) and of some marine coelenterate

Tab. 1 : Types of the host-symbiont-interface in symbiotic systems with algae[1]

Alga	Host	Type of symbiotic system: ectobiotic	endobiotic	endocytobiotic
cw	cw	Lichens	*Azolla*	-
cw	mb	Lichens	Molluscs	Ciliates, *Hydra*, Corals
cw	cp	-	-	-
mb	cw	-	-	-
mb	mb	-	*Convoluta* ?	*Glaucosphaera* ?
mb	cp	-	-	symbiotic Chloroplasts?, *Peridinium balticum*, Dinoflagellates ?

1 : selection, cw : cell wall, mb : plasma membrane, cp : cytoplasm, - : not observed

species with dinoflagellate algae (Fitt and Trench 1983).
In endocytobiotic systems, the formation of a stable host-symbiont-interface requires the development of mechanisms which allow the uptake and maintenance of a genetically different partner organism in a host cell. Thus it is interesting that in all endocytobiotic systems with algal partners which have been studied so far, that problem is mastered by similar strategies:

(a) Algae are taken up by host cells which are capable of phagocytosis and hence can form lysosomal compartments.

(b) Algae are enclosed in special compartments of host cells (so-called "Perialgal Vacuoles" or "Symbiosomes") where they are protected against the attack of host lytic enzymes.

(c) The individual enclosure of algae in perialgal vacuoles guarantees a maximum host-symbiont-interface area and high capacity for exchange of signals.

(d) The enclosure of algae in special compartments is specific, i. e. only a limited number of algal strains is able to form a stable symbiotic system.

(e) When algae divide in perialgal vacuoles, daughter cells are immediately separated from each other and enclosed again individually in new perialgal vacuoles.

Thus it can be concluded that the formation of a specific host-symbiont-interface

(i) results from a cell-to-cell recognition process between host and algae which does not only take place during phagocytosis of algae , i. e. when a new symbiotic relationship is established, but also when already established algal partners divide in a host cell, and which

(ii) is achieved in different endocytobiotic systems in a similar way.

2. The Green *Paramecium* System

The green *Paramecium* is a hereditary endocytobiotic association of *Paramecium bursaria* (Ciliata) and *Chlorella* sp. (Chlorophyceae) (for a review see Reisser 1986). Under natural conditions ciliates always contain algal partners. When the host cell divides, algae are statistically distributed to daughter cells. Under laboratory conditions partners can be separated from each other and grown independently as mass cultures. Thus it is possible to perform infection experiments: Exsymbiotic (isolated from the symbiotic system) or aposymbiotic (freeliving) algae can be offered to exsymbiotic or aposymbiotic ciliates which afterwards are checked by light or electron microscopy for the formation of perialgal vacuoles (for a protocol of methods see Niess et al. 1982).

Infection experiments clearly show a marked specificity: Although a broad range of different freeliving and exsymbiotic *Chlorella* strains are phagocytosed by the ciliate, only few of them are taken up into perialgal vacuoles (Fig. 1).

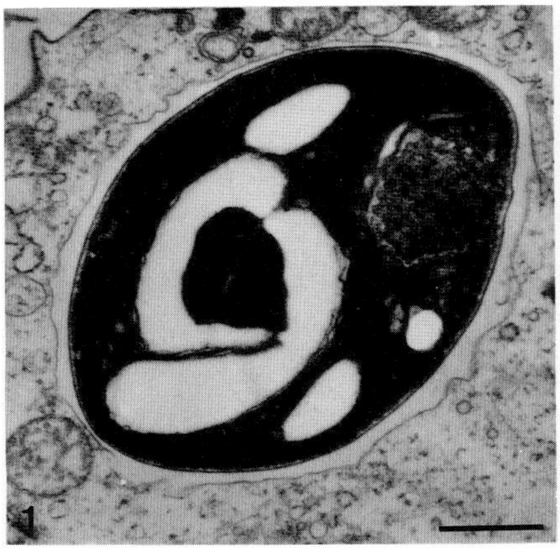

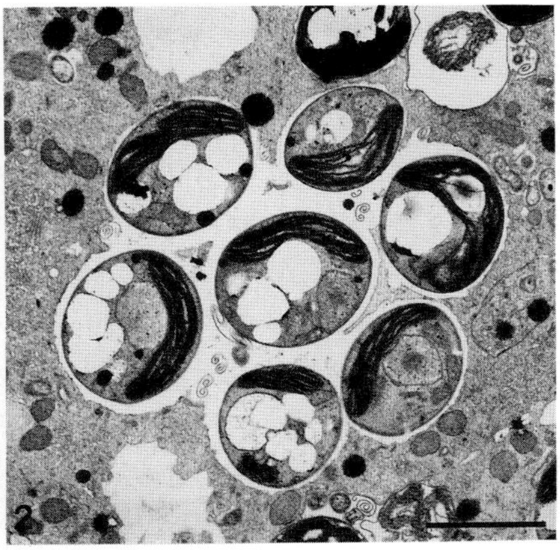

Fig. 1 : **Symbiotic *Chlorella* sp. in a perialgal vacuole of *Paramecium bursaria***

bar : 1 μm

Fig. 2 : **Shrinking digestive vacuole of *Paramecium bursaria***

Note close attachment of the vacuolar membrane to algal cell wall surfaces.
bar : 5 μm

The other ones are sequestered into digestive vacuoles (Fig. 2). Obviously only few algal strains are recognized by the ciliate as suitable for symbiosis formation, i. e. are able to trigger a positive recognition process, which leads to the formation of a stable host-symbiont-interface and an enclosure into perialgal vacuoles. It then depends on the physiological adaptation to their symbiotic habitat whether algae populate their host permanently or are lost again after a certain amount of host division steps. That phenomenon of specificity is a common feature of stable symbiotic associations and has also been reported for the green *Hydra*- and for dinoflagellate-coral-systems (Fitt and Trench 1983, McNeil and McAuley 1984).

The host-symbiont-interface of the green *Paramecium* system can be characterized by both ultrastructural and physiological features (for an overview see Reisser 1988) : The perialgal vacuole membrane differs from membranes of digestive vacuoles by size and distribution of intramembranous particles. Primary and secondary lysosomes do not fuse with it. The exchange of metabolites (O_2, CO_2, sugars, amino acids) between partners is mediated through the perialgal vacuole membrane. It probably plays also an important role as a part of the sensory transduction chain between alga and ciliate which is required for the light dependent swimming behaviour of the symbiotic unit.

2. 1. Formation Of The Host-Symbiont-Interface : The Alga

A lot of effort in symbiosis research has centered on recognition processes between partners, i. e. on the question how a stable host-symbiont-interface is formed. It is an accepted working hypothesis that the formation of a stable interface depends on special algal characters although the problem how the putative host is able to recognize those characters is usually ignored.

Special characters of algae can be either physiological or structural features. It is well known that symbiotic *Chlorella* strains isolated from different hosts such as ciliates , *Hydra viridis* or green sponges excrete sugars (xylose, glucose, maltose, trehalose) yet the signal character of those sugars for recognition events could never be proved (Fischer et al. 1989, Pardy 1989, Muscatine et al. 1967, Reisser et al. 1982). In the green *Paramecium* system, it could be shown definitively that also algae which do not excrete sugars can trigger the formation of a perialgal vacuole, i. e. are recognized as suitable for symbiosis formation. Sugar excretion of algae in this system is of adaptive value only for the establishment of a long lasting and stable hereditary symbiotic relationship in the natural environment (Reisser 1987).

Experiments clearly document the importance of algal cell wall surface structures for the recognition process: As is illustrated by Tab. 2, the chemical composition of cell walls of exsymbiotic and aposymbiotic algae is different. Every time the cell wall of the exsymbiotic *Chlorella* is changed, either by cell wall degrading enzymes or by the coupling of lectins or antibodies, it is no more recognized by the ciliate as suitable for symbiosis formation, i. e. is no more able to trigger the formation of a stable host-symbiont-interface. Thus a screening programme was started in order to compare different features of cell wall surfaces of both exsymbiotic and aposymbiotic chlorellae for structures which may be required to trigger the formation of a stable host-symbiont-interface. It could be shown that cell wall surfaces of different *Chlorella* species and strains differ from each other by surface charge and binding capacity of different lectins and antibodies (Reisser 1988, Reisser et al. 1982). Binding of different lectins gives some hints to a special pattern common to chlorellae which induce a stable host-symbiont-interface (Reisser 1988). Probably inducing algae are recognized by the ciliate by a sort of fingerprint-like recognition pattern of carbohydrates on their cell wall surfaces.

In the association of *Hydra viridis* and *Chlorella* sp. studies with antibodies (Pool 1979) and lectins (Meints and Pardy 1980) also suggest a carbohydrate recognition pattern on the cell wall surfaces of algae. Dinoflagellates isolated from different marine hosts differ in cell wall surface patterns as could be shown by antibody techniques (Kinzie and Chee 1982).

However, it should be stressed that by lectins or antibodies only the existence of common structural patterns on the surfaces of inducing algae can be detected. There is no way to be sure that those structures in fact are the trigger for the formation of a stable host-symbiont-interface: It could be shown repeatedly that coupling of every type of lectins or antibodies to the surface of a symbiotic alga blocks a positive recognition process and thus results in a digestion of the alga (Reisser et al. 1982). Thus it cannot be excluded that lectins or antibodies which are coupled to the surfaces of inducing algae can block by sterical hindrance other cell surface components which might be the

Tab. 2 : Binding of a lectin[a] to the cell wall surfaces of exsymbiotic and aposymbiotic *Chlorella* spp. after treatment with cellulase or pectinase

	lectin bound[b]		infection[c]	
Treatment with	symb. *C.*	apos. *C.*	symb. *C.*	apos. *C.*
Control A	–		+	–
Control B	100	100	–	–
Cellulase	116	161	–	–
Pectinase	83	137	–	–

a : Concanavalin A, b : relative units per standard surface area after 60 min enzyme treatment, +/- 5 %, c : infection of exsymbiotic *Paramecium bursaria*, symb. *C.* : symbiotic *Chlorella* sp. exsymbiotic from *Paramecium bursaria* , apos. *C.* : aposymbiotic *Chlorella vulgaris* var. *vulgaris*

"true" recognition groups. In the green *Paramecium*, recognition mechanisms are so specific that the ciliate is able to discriminate autospores and mother cells of algae : Infection experiments with a *Chlorella* sp. exsymbiotic from *Paramecium bursaria* and cell-cycle specific antibodies to it (Reisser et al. 1982), show that only autospores are enclosed in perialgal vacuoles. Mother cells are digested. Differences in cell wall surface patterns of autospores and mother cells are also documented by amounts of lectins bound to them (Tab. 3).

2. 2. Formation Of The Host-Symbiont-Interface : The Host

While a lot of effort has been spent on the study of putative recognition groups of the algal partner, virtually no information was available on mechanisms by which the host could be able to recognize those groups. However, it is obvious that a special host recognition system must exist, since among different species of a host genus usually only one species is able to form a symbiotic relationship (Reisser 1986). Recent studies on recognition mechanisms in the green *Paramecium* system prove the existence of a factor in the *Paramecium* cell which binds to algal surfaces. That factor was called Algal Agglutinating Factor (AAF) and could be detected not only in *Paramecium bursaria* but also in aposymbiotic *Paramecium* strains such as *Paramecium tetraurelia*, although in lesser amounts (Reisser 1989, Tab. 4).

AAF binds to the surfaces of exsymbiotic and aposymbiotic *Chlorella* strains and agglutinates algae. The highest amounts of AAF per standard algal surface area are bound to the exsymbiotic *Chlorella* strain which has been isolated from *Paramecium bursaria*. Current studies try to analyse the chemical and physical properties of AAF (for details see Reisser 1989).

Preliminary results show that AAF is a soluble protein with a molecular weight of about 30 kd which is loosely bound to ciliate membranes. Besides chlorellae it also agglutinates erythrocytes, and its agglutinating activity can be blocked in the presence of galactose. Therefore it may be speculated that AAF has lectin-type features.

3. Mechanisms Of Host-Partner-Recognition : The Zipper-Model

From above data a new model of cell-to-cell recognition mechanisms in the green *Paramecium* system has been developed which is illustrated by Fig. 3. According to that model, recognition between putative partners bases on an interaction of a ciliate lectin and algal cell wall surface bound carbohydrates. That interaction works like a zipper mechanism and thus leads to the enclosure of a suitable alga by the host membrane. Only those algae are enclosed which have a sufficient amount of suitable carbohydrates on their surfaces to start the zipper mechanism. If that amount is too small - as it is the case in non-inducing algae - cells can not be

Tab. 3 : Infection of exsymbiotic *Paramecium bursaria* with autospores and mother cells of symbiotic *Chlorella* sp. isolated from green *Paramecium bursaria* and treated with different antibodies[a]

Infection with	untreated algae[b]	algae treated with control serum	algae treated with antiserum to autosp.	algae treated with antiserum to mother c.
autospores	85	80	20	75
mother cells	5	5	5	5

a : numbers give % of permanently infected paramecia,
b : untreated mother cells bind about 100 % more Concanavalin A per standard surface area than autospores

enclosed, recognition is negative. That model also gives an explanation why *Paramecium* species other than *Paramecium bursaria* don`t form symbiotic associations with Chlorella sp. Their amount of membrane bound AAF is much smaller than in *Paramecium bursaria* (Tab. 4) and probably therefore an individual enclosure of algae is not possible.

While a lot of studies on mechanisms of cell recognition among higher plant cells and between higher plant cells and prokaryotes have been published (for an overview see Reisert 1981, Sharon and Lis 1989) only few data are available on the molecular basis of interactions between symbiotic algae and host cells. In lichens, lectins are discussed to be involved in recognition events (Galun and Bubrick 1984). Corresponding data on algae-invertebrate associations are lacking so far. Basing on experiments and data reported in this paper a new mechanism of cell-to-cell recognition in endocytobiotic systems of invertebrates and algae is proposed which shall be called "zipper mechanism". This term has been coined for the first time by Griffin et al. (1976) in describing phagocytosis of macrophages.

Besides data reported here also other ultrastructural and biochemical studies on the green *Paramecium* system support the

Tab. 4 : Algal agglutinating activity of the ciliate-fraction prepared from green *Paramecium bursaria*

Agglutination of	minimum amount of protein needed for agglutination[a,b]
symbiotic *Chlorella* spp.	
isolated from ciliates	
Paramecium bursaria[c]	1.00
Euplotes daidaleos	4.31
isolated from lichen	
Trapelia coarctata	
(3.80)[d]	5.31
aposymbiotic *Chlorella*	
C. vulgaris (211-11b)[d]	3.56
Erythrocytes (Sheep)	0.16

a : (rel. units × standard cell surface area^{-1}),
b : agglutinating activity in preparations from green *Paramecium bursaria : Paramecium tetraurelia* = 3 : 1;
c : recognition positive, enclosure in perialgal vacuole;
d : strain number of Göttingen Algal Collection

zipper-model of recognition:
(a) Algae which are phagocytosed by the *Paramecium* are enclosed together with bacteria in digestive vacuoles. Those vacuoles start cyclosis by fusion with acidosomes and then pass through a shrinking process before fusion with lysosomes occurs (for cyclosis events in *Paramecium* spp. see Allen 1984). It could be shown that perialgal vacuoles are formed shortly after acidification of the vacuole by acidosomes but before lysosomes fuse and digestion of the vacuolar contents starts (Meier and Wiessner 1989). Thus it is conceivable that by shrinking of the vacuole a maximum surface contact between the vacuolar membrane and the enclosed algae and thus recognition events are rendered possible (Fig. 2). Such an enclosure of a symbiotic *Chlorella* into a separate vacuole of *Paramecium bursaria* has also been documented in a

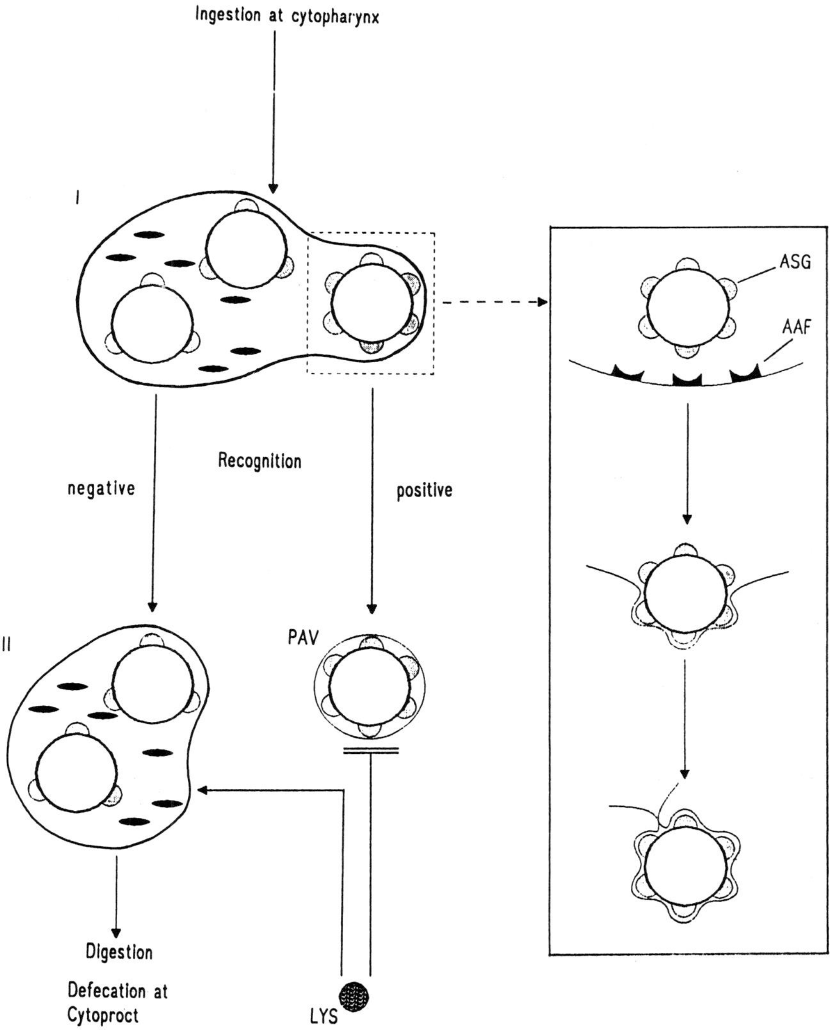

Fig. 3 : **Cell-to-cell recognition in the *Paramecium bursaria-Chlorella* sp. association ("zipper-model")**

I : shrinking digestive vacuole enclosing algae (circles) suitable or not suitable for symbiosis formation and bacteria (disks), II : digestive vacuole of a later stage of cyclosis, PAV : Peri-Algal Vacuole, Lys : Lysosome, AAF : Algal Agglutinating Factor, ASG : Algal Surface Group

series of electron micrographs by Meier and Wiessner (1989).
(b) It could be shown that the perialgal vacuole membrane does not only differ significantly from the digestive vacuole membrane by size and distribution of intramembraneous particles but that it is also rather rigid in structure: While the membrane composition of digestive vacuoles tremendously changes during cyclosis because of a lot of fusion and pinching off events, comparative changes have not been observed in perialgal vacuole membranes of *Paramecium bursaria* (Meier et al. 1984). This "conservatism" is possibly due to the fact that the perialgal vacuole membrane is closely attached to the algal cell wall surface because of the lectin-type binding mechanism. It is conceivable that such a close attachement decreases the fluidity of the perialgal vacuole membrane to such a degree that any fusion or pinching off by vesicles is not possible any more. Thus fusion of perialgal vacuoles with primary or secondary lysosomes has never been observed and any digestive activity could not be shown in perialgal vacuoles (Karakashian and Karakashian 1973, Reisser et al. 1983).
Obviously also in other invertebrate-algae-systems a zipper-like recognition mechanism is conceivable: As could be shown by electron microscopic studies on infection of *Chlorohydra viridissima* by an exsymbiotic *Chlorella* sp., membranes of host phagocytic cells attach to algal cell wall surfaces and close around them (McNeil 1981). A similar mechanism of attachement and enclosure has been observed during infection of *Cassiopeia xamachana* (Coelenterata) with *Symbiodinium microadriaticum* (Dinophyceae) (Fitt and Trench 1983).
The "zipper-model" presented here is the first one to suggest a lectin based recognition mechanism in endocytobiotic systems. What is more, data also suggest that in the green *Paramecium* recognition bases rather on quantitative than on qualitative criteria : The formation of a stable host-symbiont-interface by the zipper mechanism is only possible when a certain critical amount of algal surface groups is present. Inducing and non-inducing algae do not differ in the ability to form recognition groups but in the amount of groups formed by them.

4. Acknowledgements

Part of experiments reported here are supported by a grant of the Deutsche Forschungsgemeinschaft to the author.

5. References

Allen RD (1984) *Paramecium* phagosome membrane: from oral region to cytoproct and back again. J Protozool 31:1-6

Fischer A, Meindl D, Loos E (1989) Glucose excretion by the symbiotic *Chlorella* of *Spongilla fluviatilis*. Planta 179:251-256

Fitt WK, Trench RK (1983) Endocytosis of the symbiotic dinoflagellate *Symbiodinium microadriaticum* Freudenthal by endodermal cells of the scyphistomae of *Cassiopeia xamachana* and resistance of the algae to host digestion. J Cell Sci 64:195-212

Galun M, Bubrick P (1984) Physiological interactions between the partners of the lichen symbiosis. In: Encyclopedia of Plant Physiology, New Series 17, Intercellular Interactions, Linskens HF, Heslop-Harrison J (eds) 362-401, Springer Berlin Heidelberg New York

Griffin FM, Griffin JA, Silverstein SC (1976) Studies on the mechanism of phagocytosis. II. The interaction of macrophages with anti-immunoglobulin IgG-coated bone marrow-derived lymphocytes. J Expl Med 144:788-809

Karakashian MW, Karakashian SJ (1973) Intracellular digestion and symbiosis in *Paramecium bursaria*. Expl Cell Res 81: 111-119

Kinzie RA, Chee GS (1982) Strain-specific differences in surface antigens of symbiotic algae. Appl Environ Microbiol 44:1238-1240

McNeil PL (1981) Mechanisms of nutritive endocytosis. I. Phagocytotic versatility and cellular recognition in *Chlorohydra* digestive cells, a scanning electron microscopic study. J Cell Sci 49:311-339

McNeil PL, McAuley PJ (1984) Lysosomes fuse with one half of alga-bearing phagosomes during the reestablishment of the european green *Hydra* symbiosis J Expl Zool 230:377-385

Meier R, Wiessner W (1989) Infection of alga-free *Paramecium bursaria* with symbiotic *Chlorella* sp. isolated from green paramecia. II. A timed study. J Cell Sci 93: 571-579

Meier R, Lefort-Tran M, Pouphile M, Reisser W, Wiessner W (1984) Comparative freeze-fracture study of perialgal and digestive vacuoles in *Paramecium bursaria*. J Cell Sci 71: 121-140

Meints RH, Pardy RL (1980) Quantitative demonstration of cell surface involvement in a plant-animal symbiosis: Lectin

inhibition of reassociation. J Cell Sci 43:239-251
Muscatine L, Karakashian SJ, Karakashian MW (1967) Soluble extracellular products of algae symbiotic with a ciliate, a sponge and a mutant *Hydra*. Comp Biochem Physiol 20:1-12
Niess D, Reisser W, Wiessner W (1982) Photobehaviour of *Paramecium bursaria* infected with different symbiotic and aposymbiotic species of *Chlorella*. Planta 156:475-480
Pardy RL, Spargo B, Crowe JH (1989) Release of trehalose by symbiotic algae. Symbiosis 7:149-158
Pool RR (1979) The role of algal antigenic determinants in the recognition of potential algal symbionts by cells of *Chlorohydra*. J Cell Sci 35:367-379
Reisert PS (181) Plant cell surface structure and recognition phenomena with reference to symbiosis. Int Rev Cytol Suppl 12:71-112
Reisser W (1986) Endosymbiotic associations of freshwater protozoa and algae. Progr Protozool 1:195-214
Reisser W (1987) Naturally occurring and artificially established associations of ciliates and algae: Models for different steps in the evolution of stable endosymbioses. Ann NY Acad Sci 503:316-329
Reisser (1988) Signals in the *Paramecium bursaria-Chlorella* sp. association. NATO ASI Ser H17:271-282
Reisser W (1989) New data on cell-to-cell recognition in the *Paramecium bursaria-Chlorella* sp.-association: Recognition is a quantitative phenomenon basing on a lectin-type mechanism. Endocytobiol 4 (in press)
Reisser W, Meier R, Kurmeier B (1983) The regulation of endosymbiotic algal population size in ciliate-algae associations. An ecological model. Endocytobiol 2:545-552
Reisser W, Radunz A, Wiessner W (1982) The participation of algal surface structures in the cell recognition process during infection of aposymbiotic *Paramecium bursaria* with symbiotic chlorellae. Cytobios 33:39-50
Reisser W, Fischer-Defoy D, Staudinger J, Schilling N, Hausmann K (1982) The endosymbiotic unit of *Climacostomum virens* and *Chlorella* sp. I. Morphological and physiological studies on the algal partner and its localization in the host cell. Protoplasma 119:93-99
Sharon N, Lis H (1989) Lectins as cell recognition molecules. Science 246:227-234

Autolysins in *Chlamydomonas*

S. Waffenschmidt and L. Jaenicke
Institut für Biochemie, Universität zu Köln
An dert Bottmühle 2, D-5000 Köln 1, FRG

Introduction

Cells of higher and lower plants are surrounded by a rigid cell wall, which is composed of elaborately intermeshed macromolecules. Growth, release and reproduction require a modification of this closely-knitted mesh-work.

In the genus *Chlamydomonas* the cell wall is built from globular and fibrillar glycoproteins rich in hydroxyproline II-helices, forming a multilayered hull around the protoplast. To open this encasement at least two defined processes of cell wall lysis are needed: (1) when daughter cells hatch from the mother cells at the end of the vegetative development; (2) when wall-less gametes of opposite mating type fuse to form the zygote in the generative cycle. Schlösser (1966, 1976) found that during these processes cell wall lytic activities were secreted into the medium which he termed "autolysins". (The generally accepted term is "lysin", and the phase specificity is denoted by "V" or "G" as explained below.) The lysins differ characteristically in the specificity of their respective substrates. The lysin of the mating pairs (gamete lysin = G-Lysin) dissolves the cell walls of gametes as well as those of all vegetative cell stages; the lysin of hatching zoospores (vegetative lysin = V-lysin) lyses the walls of zoospore mother cells (sporangia) only. In addition to this stage specificity, cross-reactivity studies with V-lysin preparations and heat-killed sporangia of different *Chlamydomonas* strains demonstrated a strain-specificity for these lytic processes. This has been used for detailed taxonomic studies (Schlösser, 1989).

This review will summarize our investigations on the biochemistry of these cell wall lytic phenomena, carried out mostly with the species *Chlamydomonas reinhardtii*, but also with *C. eugametos, C. smithii, C. mexicana* and *C. debaryana. C. reinhardtii* offers a number of advantages. (i) It is the one member of a rather mixed genus that is best understood genetically and physiologically. (ii) Conditions for synchronization of its life cycle have been well worked out (Schlösser, 1966). This is a prerequisite for obtaining uniform material for detailed biochemical analyses of either cell walls or lysins. (iii) The asexual (vegetative) life cycle can be experimentally switched to the sexual (gamete) life cycle by nitrogen starvation (Sager and Granick, 1954). (iv) The cell wall is ultrastructurally well defined (Roberts et al., 1972, Goodenough and Heuser, 1985), and molecular data of some of the wall s main components are at hand (Miller, 1972; Catt et al., 1976; Adair and Apt, 1989). In contrast to the polysaccharide-containing cell walls of higher plants the *Chlamydomonas* cell wall is predominantly built of more or less salt-soluble, globular or fibrous hydroxyproline-rich glycoproteins, stacked in 5 discernible layers numbered W_1, W_2, W_4, W_6 and W_7 from the periplasm to the surface (Goodenough and Heuser, 1985; Adair, 1988). The glycoproteins of the layers W4 and W6 can be extracted by chaotropic reagents such as LiCl or $NaClO_4$ (Catt et al., 1976). The inner wall layer is highly cross-linked; therefore little is known of the existence and nature of cross-linking bonds. However, according to electronmicroscopic observations it is this part of the cell wall which is nicked and split by the lytic enzymes (Goodenough and Heuser, 1985).

It is the **stage-specific** action of the lytic processes that attracted our special interest; therefore, most of the data present here emphasize the molecular basis of this aspect.

Biochemical characterization of V- and G-lysin

The "lysing principles" described by Schlösser are defined enzymatic processes, catalysed by specific proteases (Jaenicke et al., 1987). Isolation and characterization of these enzymes require reliable assays of their action. V-lysin activity is followed by staining sporangia with the polysaccharide specific dye Remazol Brilliant Blue and measuring spectrophotometrically the release of blue coloured (λ_{max} = 605 nm) fragments from the cell wall upon enzyme action (Jaenicke et al., 1987). Under standard conditions (fixed number of stained cells; defined incubation time and temperature) this convinient assay is linear with V-lysin activity. G-lysin produces protoplasts of

Table 1 : Properties of Lysins in *Chlamydomonas reinhardtii*

	V-lysin	G-lysin
Molecular mass deglycosylated	34 kDa 30 kDa	65 kDa 56 kDa
Quaternary structure	single chain	single chain
Substrate wall	sporangia	gametes, zoospores, sporangia
Other substrates	azocoll gelatine	azocoll gelatine
Inhibitors	EDTA (2 mM)	EDTA (0.4 mM), diisopropylfluorophosphate (1 mM), diethylpyrocarbonate (0.5 mM)
Bound metal	Zn^{2+}, Mn^{2+} (3 : 1)	no metal ion found

vegetative cells and gametes, which are labile against detergents. Incubation of a given number of zoospores, lysis of the wall-less cells with 0.05 % Triton X-100 and measurement of the released chlorophyll (λ = 440 nm) gives an assay which is proportional to incubation time and enzyme concentration when standardized. Using these assays the V- and G-lysin of *C. reinhardtii* were purified from hatching or mating media, respectively, by conventional protein purification steps such as ammonium sulphate or acetone precipitation, ion exchange, size exclusion and hydrophobic interaction chromatography (Jaenicke et al., 1987). V- and G-lysin are two distinct enzymes of different molecular size but with similar characteristics and catalytic properties, as summarized in Table 1.

The hydrolytic activity depends on the presence of divalent cations, particulary Ca^{2+} and can be inhibited by chelators such as EDTA and EGTA. Besides their specific *C. reinhardtii* cell wall substrate, both enzymes hydrolyze related (hydroxy)proline containing matrix proteins, such azocoll, gelatine and heat-denaturated collagen, but not general proteinase substrates such as casein or hemoglobin and small hydroxyproline peptides. Analysis by atomic emission spectroscopy of the purified and exhaustively dialyzed proteins indicated that V-lysin contains bound Zn^{2+} and Mn^{2+} in a ratio 3 : 1. However, contrary to the report of Matsuda and coworkers (Matsuda et al., 1985), neither Zn^{2+} nor any other divalent cation is found in G-lysin. SH-groups modifying reagents in mM concentrations are without effect on both enzymes, but inhibitors esterifying serine or alkylating histidine block the activity of G-lysin. Ongoing work to clone and sequence the genes of both V- and G-lysins will enable a comparison with other proteases and will help to elucidate the hydrolytic mechanisms of the two enzymes.

Characterization of autolysins during the *Chlamydomonas* cell cycle

Our finding that *Chlamydomonas* produces two different lysins in its generative cycles was the first step in elucidating the molecular basis of the strategy of stage-specific cell wall lysis. A logical next step was to determine the amount and formation of both enzymes during asexual and sexual reproduction. G-lysin is secreted into the medium during mating of gamtes of opposite mating type, suggesting that either the expression of this protein or its release from stored forms is increased during gametogenesis. The specific activity of G-lysins is much higher in homogenates of gametes than in that of zoospores. To compare the relative amount of G-lysin protein in these homogenates we put to use the observation that G-lysin is resistant to tryptic cleavage. If tryptic digests of cell homogenates of different stages are separated by SDS-PAGE, the patterns shown in Fig. 1 were observed: Most of the cell proteins of gametes (lane a) and zoospores (lane c) are affected, but a 65 kDa protein is left intact in the digest of gamete homogenates (lane b). This protein is obviously not present in the digest of zoospore homogenates (lane c).

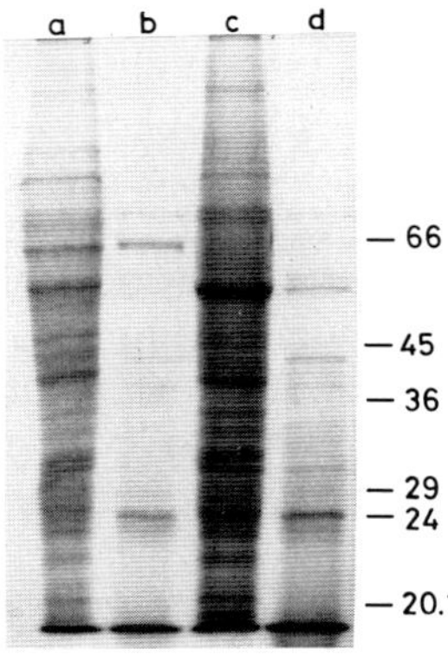

Fig. 1: Cell homogenates of zoospores and gametes (1 mg protein) were incubated with 50 ug of trypsin for 60 min and then separated on a 10 % SDS-PAGE and stainedwith Coomassie Brilliant Blue

With polyclonal antibodies raised in rabbits against G-lysin we could demonstrate by immunoblot analysis that the 65 kDa protein is identical with G-lysin and that the amount of G-lysin is increased during gamete formation (Waffenschmidt et al., 1989). When gametogenesis was induced in vegetative cultures by nitrogen starvation, homogenates of the cells showed increasing G-lysin activity parallel to the formation of the 65 kDa G-lysin. Fig. 2 shows the time course of G-lysin activity (A), the protein pattern after SDS-PAGE (B) and the corresponding immunoblots of cell homogenates during gamete differentiation (C).

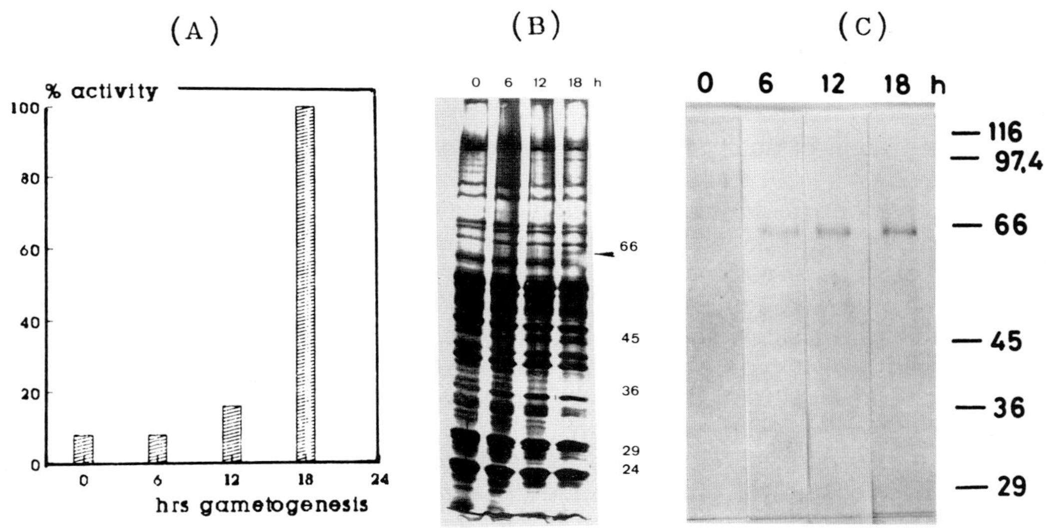

Fig. 2 : Cells grown in suspension cultures were transferred to N-free medium At the times indicated portions were sonicated. In the cell homogenates the following parameters were determined : lysin activity (A), protein by SDS-PAGE (B) and immunoblotting with G-lysin antibodies (C).

W. Snell, confirming Y. Matsuda, found that G-lysin is synthesized as an inactive zymogen and stored in the periplasm in which it is activated by proteolytic cleavage by a "prolysinase" (Matsuda et al., 1987; Buchanan et al., 1989). This activation occurs *in vitro* by addition of a cAMP-analogue (Pasquale and Goodenough, 1987). Immunohistological studies are in progress in cooperation with M. Melkonian to localize G-lysin within the gamete cells. As first preliminary results we found the Golgi apparatus as the site of G-lysin processing from which the enzyme is subsequently transported to the periplamn. This is consistent with immunoblot analyses and activity assays of the cell fractions from sucrose gradient centrifugations. G-lysin activity and antigen are predominant in the microsomal preparations derived from ER and Golgi membranes.

The time of the synthesis of V-lysin during vegetative life cycle of *C. reinhardtii* was examined by a similar approach. Antibodies raised in rabbits against the native enzyme were purified from unspecific antibodies binding to common carbohydrate epitopes by adsorption to isolated inner wall fragments. They proved highly specific to trace the protein immunologically. Secretion of the enzyme protein was followed by analyzing aliquots of cell free medium for lytic activity and immunoreactive protein. Both increase just at the time when the daughter cells hatch. We infer from these experiments that activation and secretion of the V-lysin occurs simultaneously and is highly controlled. Efforts to trigger both processes *in vivo* with cAMP, its analogues, or drugs interfering with cAMP degradation have failed. Light has some effects on the lysis of mature sporangia. However, it can not be the primary signal, since immature sporangia, which contain V-lysin as shown immunologically, do not release their zoospores upon exposure by light, but proceed, instead, with mitotic divisions. Only when sporangia have matured they do release their progeny at the onset of illumination. We have also studied V-lysin formation during the vegetative cell cycle by analyzing homogenates from synchronously grown cultures by SDS-Page, by immunoblotting, and by spectrophotometric assays for enzy-

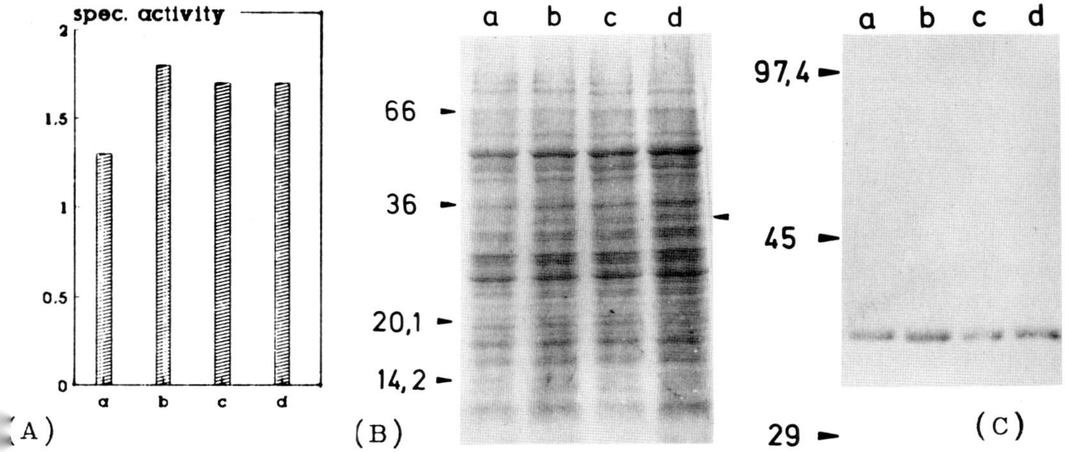

Fig. 3 : Cell homognates of synchronously grown suspension cultures were assayed for enzyme activity (A), for protein by SDS-PAGE (B) and by immunoblotting with V-lysin antibodies (C)

me activities. When cells resorb their flagella before the first mitotic division, cell wall lytic activity rises (Fig.3A) in parallel with an anti V-lysin immuno-reactive protein of 34 kDa, the molecular mass of V-lysin (Fig. 3B and 3C). In accordance with our results regarding the G-lysin Y.H. Zhang (1989) using our antibodies has identified the endomembrane fractions ER and Golgi as being the sites of V-lysin synthesis and processing.

Additional evidence for the stage-specificity of V-lysin synthesis was obtained by translating isolated mRNAs from defined cell stages in a rabbit reticulocyte system (^{35}S-methionine as labelled amino acid) and precipitating the translation products with the anti-V-lysin adsorbed to protein A-Sepharose. A 34 kDa protein was precipitated with the same time course as V-lysin in Fig. 3. Since the deglycosylated V-lysin has a molecular mass of only 30 kDa, the 4 kDa difference might be due to a "pre-V-lysin" carrying the additional N-terminal signal peptide common for secretory proteins (Fig. 4).

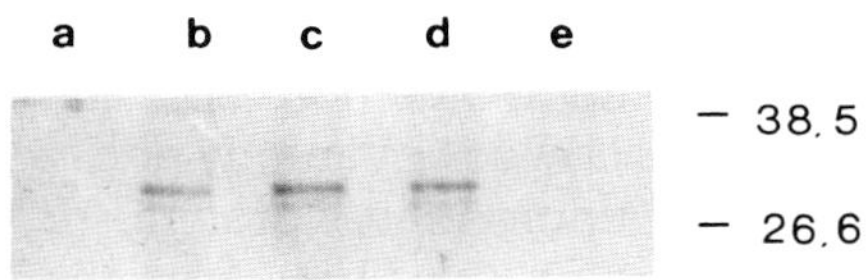

Fig. 4 : Total mRNA from different cell stages were isolated and translated in a reticulocyte lysate with ^{35}S-methionine. ProteinA-sepharose coated with anti-V-lysin were used to specifically immunoprecipitate translational products

Specificity towards wall substrates

A specificity of autolysins towards their cell wall substrate molecule(s) is to be expected in addition to stage-specific activation and secretion. Obvious questions as to the specificity of the lysing enzymes are therefore: Which of the constituents do they attack? Do they hydrolyze glycosidic or peptidic bonds? Is there a preferred splitting sequence? Can the specificity of G-lysin and V-lysin be explained by chemical differences between mother and daughter cell walls, or is it due to enzyme kinetic parameters of ligand interactions?

We have fractionated cell walls and purified their components in order to find a defined substrate molecule for the lysins. Isolated cell walls of *C. reinhardtii* were exrtracted with neutral $NaClO_4$ and with mildly acidic oxidizing sodium chlorite, after which an insoluble portion remained (Jaenicke et al., 1987). The three fractions (PSW = perchlorate soluble wall, CSW = chlorite soluble wall and ISW = insoluble cell wall) were incubated with the lysins, and the protein pattern of the hydrolysates by SDS-PAGE compared with untreated controls. The main glycoproteins (GP1, GP2 and GP3) of PSW were left intact, whereas highly aggregated CSW and ISW gave defined cleavage products. However, most of the high-molecular wall components did not enter the stacking gel, so that is was impossible to find a defined substrate (Waffenschmidt et al., 1988). Recently we have succeded in isolating a 28 kDa subunit of the PSW by electroeluting the glycoprotein from gels after SDS-PAGE. It is split by V-lysin into four defined fragments with a molecular mass ranging between 4 and 7 kDa (SDS-PAGE). It should therefore now be possible to identify the preferred splitting sequences in this substrate. This fragmentation gives further evidence that the lysins are proteases, as has already been inferred from their action on protein substrates as shown in Table 1. Final proof that V-lysin degrades its substrates by proteolysis was obtained when we followed sporangial wall lysis by specific assays for glycosidase and protease activity in the same enzyme preparations measuring an increase of reducing sugars by the biscinchoninate method (Waffenschmidt and Jaenicke, 1987), and by increase of free amino groups by the fluorescamine method (Schwabe, 1973). Whereas there was no rise in reducing sugars, fluorescence increased with time due to the liberation of N-terminals as a result of the action of V-lysin on sporangial wall CSW (Fig. 5).

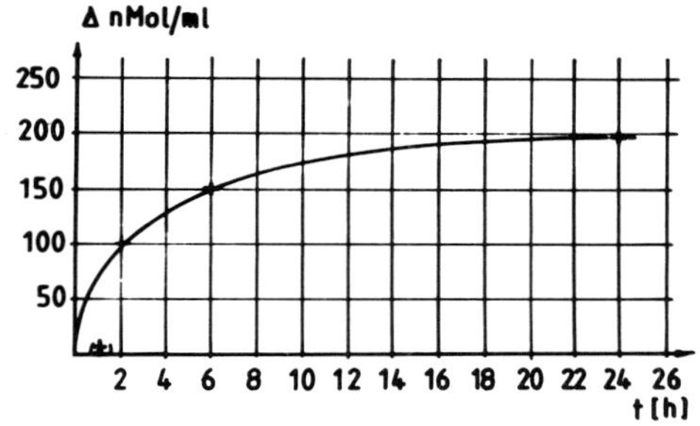

Fig. 5 : Release of amino groups as measured with the fluorescamine method on incubating sporangial CSW with V-lysin

With this method the enzyme constants V_{max} and $K_{0.5}$ of V- and G-lysin for the wall fractions (PSW, CSW and ISW) of different cell stages (sporangia, zoospores) as substrates were defined. Both enzymes hydrolyze the ISW fractions quickest and show the highest affinity towards this substrate. This is consistent with electron microscopic observations (Goodenough and Heuser, 1985; Imam et al., 1985), that this cell wall layer is lysed first on incubation of cells with crude lysin. ISW of sporangia, young zoospores and gametes were degraded without preference by G-lysin. V-lysin, however, has a much higher affinity and rate of hydrolysis for the sporangial cell wall than for any other substrate. Some degradation of daughter cell walls is also observed, but only after much extended incubation. We conclude from these experiments that the wall molecules themselves contain the information for stage specific degradation. Specifically modified substrate molecules should give information on the molecular recognition sites and structures.

The substrate molecules are glycoproteins. When the sugar moiety is removed from wall fractions by deglycosylation with trifluoromethane sulphonic acid the lysins show altered enzyme kinetic behaviour. An shown in Fig. 6A, deglycosylation decreases to less than one tenth the affinity of V-lysins towards ISW of sporangia. According to Fig. 6B this fraction has obviously lost the information of stage specific degradation activity. (Waffenschmidt et al., 1988).

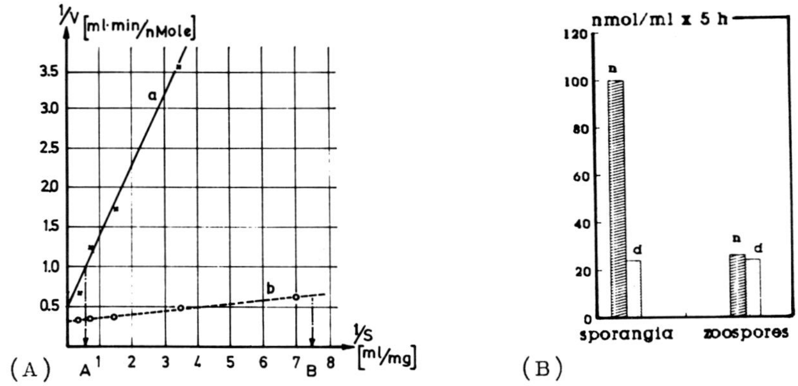

Fig. 6 : Cleavage of native (a or n) and deglycosylated (b or d) cell wall fractions by V-lysins. Lineweaver-Burk diagram for ISW (A); cleavage rates for ISW of sporangia and zoospore cell walls (B)

Table 2 summarizes the affinity constants of V-lysin and G-lysin towards native and deglycosylated ISW fractions. We conclude that the affinity of G-lysin and its substrates is determined by protein-protein interactions, whereas that of V-lysin makes use of the sugar moiety of the substrate in addition. This points to a fixed lectin-like sugar-glycoprotein interaction between the enzyme and its substrate.

Table 2: Enzyme Constants of cell wall lysis in *Chlamydomonas reinhardtii*

Substrate	V-lysin		G-lysin	
	K 0.5 ug/ml	Vmax nmol/ml x min	K 0.5 ug/ml	Vmax nmol/ml x min
ICW native				
of zoospores	2 000	-	300	3
sporangia	130	4	300	3
gametes	-	-	300	3
ICW deglycosylated				
of zoospores	-	-	300	10
sporangia	2 000	2	300	10

- = not determined

To test this hypothesis we isolated hydroxyproline oligosaccharides from the wall fractions after $Ba(OH)_2$-hydrolysis. The neutralized and partly purified (filtration on Sephadex G-25) digest was used as an inhibitor of the lysins. The enzymes were pretreated with the digest, then incubated with their appropriate ISW fraction. The fragments in digests of sporangial walls acted as a competitive inhibitor for V-lysin, indicating that sugar is essential for enzyme activity (Waffenschmidt et al. 1988). Unexpectedly fragments from digests of zoospore cell wall competitively inhibited G-lysin only with respect to the native ISW substrate and not with chemically deglycosylated ISW which inhibited in a non.-competitive fashion. Thus, although G-lysin binding is determined by a protein-protein interaction, the enzyme structure appears to be changed upon contact with an oligosaccharide so that its activity is decreased. This may provide a mechanism to position and modulate the enzyme in its biological environment.

The need for lectin-like interactions between V-lysin and its substrate could explain control of stage-specifity. Hydroxyproline oligosaccharide fragments of $Ba(OH)_2$-hydrolysates of mother and daughter cell walls were fractionated by ion exchange chromatography and HPLC. The inhibitory effect was restricted to a fraction which upon GC analysis was a mixture of the hydroxyproline-disaccharides Hyp $(Ara)_2$ and Hyp (Ara-Gal). The latter seems to be responsible for lysin inhibition, since incubation of substrate cells either with galactose-binding lectins or with ß-galactosidase and galactose oxidase make them completely inaccessible to V-lysin. Furthermore a quantitative comparison of hydroxyproline oligisaccharides within cell walls of mother and daughter cells only show a significant difference in terms of Hyp (Ara-Gal), which is in sporangia about 2fold higher than in zoospores (s. Table 3).

Table 3: Oligosaccvharides in zoospores and sporangia cell walls of *Chlamydomonas reinhardtii*

	zoospores		sporangia	
	nmol/ g cell wall	% (x 10^{-3})	nmol/g cell wall	% (x 10^{-3})
Hyp (Ara)	110.3	2.9	142.6	3.8
Hyp (Gal)	32.1	0.94	39.8	1.2
Hyp $(Ara)_2$	168.4	6.7	194.8	7.7
Hyp (Ara-Gal)	162.0	6.9	256.6	11
Hyp $(Ara_2$-Gal)	431.3	24	419.7	23
Hyp $(Ara_3$-Gal)	92.2	6.4	90.4	6.2

From these results we propose the following working hypothesis for stage specific cell wall lysis: In daughter cell walls of zoospores the attachment sites for V-lysin are masked. On maturation the binding sugar sites are created by the action of internal or integrated glycosidases or glycosyltransferases. V-lysin, after proper activation, can now bind to and act on the thus modified sporangial wall (Waffenschmidt et al. 1988). G-lysin binding is regulated not only by such term-specific processing, but rather by changes in ion fluxes and *de novo* synthesis.

Summary

Chlamydomonas reinhardtii has two proteases responsible for stage-specific cell wall lysis. V-lysis nicks the cell wall of mature zoosporangia in order to release the daughter cells. G-lysin takes off the gamete wall as a prerequisite to mating. Both enzymes are stage-specifically synthesized, (probably) stored as inactive precursors and activated by a certain signal and secreted into the medium. The glycoproteinaceous wall molecules representing the substrates contain the information for attachment and splitting (i) in the genetically fixed amino acid sequence; (ii) in the oligosaccharide side chain which may be modified to regulate stage specific splitting. This is illustrated for V-lysin in Fig. 7. Strain specificity may also reside in alterations of species-specific glycosylations of cell wall molecules.

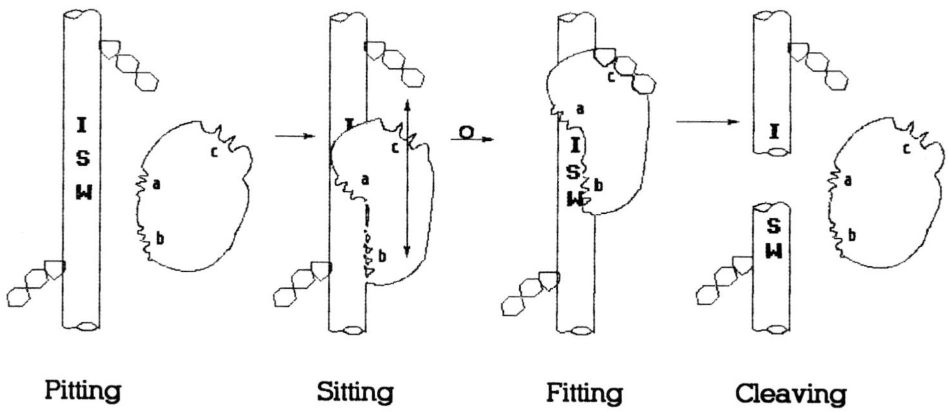

Fig. 7: Model of interactions during cell wall degradation by V-lysin.

By an essentially similar but reciprocal mechanism the activity of G-lysin might be increased allosterically by removing regulatory oligosaccharides from cell walls of zoosporangia and gametes.

Acknowledgements

The authors wish to thank a group of enthusiastic coworkers: Eva Glees, Wolfgang Kuhne, Rainer Spessert and Michael Trimborn, who helped to accumulate the data presented here.

References

W.S. Adair (1988) Organization and assembly of the *Chlamydomonas* cell wall. In: J.E. Varner, Self assembly architecture. 25-41, New York

W.S. Adair and K. Apt (1989) Cell wall regeneration in *Chlamydomonas*:Accumulation of mRNAs encoding cell wall HRPGs. Proc. Natl. Acad. Sci. USA in press

M.J. Buchanan, S.H. Imam, W.A. Eskue and W.J. Snell (1989) Activation of the cell wall degrading protease, lysin, during sexual signalling in *Chlamydomonas*: The enzyme is stored as an inactive, higher molecular mass precursor in the periplasm. J. Cell Biol. 108:199-207

J.W. Catt, G.J. Hills and K. Roberts (1976) A structural protein, containing hydroxyproline, isolated from the cell wall of *Chlamydomonas reinhardtii*. Planta 131:165-171

U.G. Goodenough and J.E. Heuser (1985) The *Chlamydomonas* cell wall and its constituent glycoproteins analysed by the quick-freeze deep-etch technique. J. Cell Biol. 101:1551-1568

S.H. Imam, M.J. Buchanan, C.-H. Shin and W.J.Snell (1985) The *Chlamydomonas* cell wall: Characterization of the wall framework. J. Cell Biol. 101:1599-1607

L. Jaenicke, W. Kuhne, R. Spessert, U. Wahle, S. Waffenschmidt (1987) Cell wall lytic enzymes (autolysins) of *Chlamydomonas reinhardtii* are (hydroxyproline)specific proteases. Eur. J. Biochem. 170:485-491

Y. Matsuda, T. Saito, T. Yamaguchi and H. Kawase (1985) Cell wall lytic enzyme released by mating gametes of *Chlamydomonas reinhardtii* is a metalloprotease and digests the sodium-perchlorate-insoluble components of cell wall. J. Biol. Chem 260:6373-6377

Y. Matsuda, T. Saito T. Yamaguchi, M. Koseki and K. Hayashi (1987) Topography of cell wall lytic enzyme in *Chlamydomonas reinhardtii*: form and location of stored enzyme in vegetative cells and gametes. J. Cell Biol. 104:321-330

D.H. Miller, D.T.A. Lamport and M. Miller (1972) Hydroxyproline heterooligosaccharides in *Chlamydomonas*. Science 176:918-920

S.M. Pasquale and U.G. Goodenough (1987) Cyclic AMP functions as a primary signal in gametes of *Chlamydomonas reinhardtii*. J. Cell Biol. 105:2279-2292

K. Roberts, M.R. Gay, G.J. Hills (1972) Structure, composition and morphogenesis of the cell wall of *Chlamydomomas reinhardtii* 1. Ultrastructure and preliminary chemical analysis. Ultrastruc. Res. 40:105-127

R. Sager and S. Granick (1954) Nutritional control of sexuality in *Chlamydomonas reinhardtii*. J. Gen. Phycol. 37:729-742

U.G. Schlösser (1966) Enzymatisch gesteuerte Freisetzung von Zoosporen bei *Chlamydomonas reinhardtii* Dangeard in Synchronkultur. Archiv f. Mikrobiol. 54: 129-159

U.G. Schlösser (1976) Entwicklungsstadien und sippenspezifische Zellwandautolysine bei der Freisetzung von Fortpflanzungszellen in der Gattung *Chlamydomonas*. Ber. Dtsch. Bot. Ges. 89:1-56

C. Schwabe (1973) A fluorescent assay for proteolytic enzymes. Anal. Biochem. 53: 484-490

R. Spessert and S. Waffenschmidt (1989) Studies on the vegetative autolysin during the vegetative life cycle of *Chlamydomonas reinhardtii*, Eur. J. Cell Biol. in press

S. Waffenschmidt and L. Jaenicke (1987) Assay of reducing sugars in the nanomolar range. Anal. Biochem 165: 337-340

S. Waffenschmidt, R. Spessert and L. Jaenicke (1988) Oligosaccharide side chains of wall molecules are essential for cell wall lysis in *Chlamydomonas reinhardtii*. Planta 175: 513-519

S. Waffenschmidt, W. Kuhne and L. Jaenicke (1989) Immunological characterization of gamete autolysins in *Chlamydomonas reinhardtii*. Botan. Acta 102:73-79

Y.H. Zang (1989) Doctoral Thesis, Göttingen

Genetic control of reproductive cell differentiation in *Volvox*

David L. Kirk
Department of Biology
Washington University
Saint Louis, MO 63130
USA

It is 100 years since August Weismann (1889) first suggested that *Volvox* might be the ideal organism in which to study the differentiation of germ and soma. But it was not until Richard Starr described the life history of *V. carteri* forma *nagariensis* (Fig. 1) exactly 80 years later (Starr, 1969) that material suitable for genetic analysis of germ-soma differentiation in *Volvox* became available to biologists[1]. Within a year of his description of normal development in this organism, Starr demonstrated its developmental-genetic potential by describing a number of Mendelian mutations that caused modified patterns of reproductive and/or somatic cell development (Starr, 1970). Huskey and his associates made substantial advances in the isolation and study of such mutants (e.g., Sessoms and Huskey, 1973; Huskey and Griffin, 1979, Callahan and Huskey, 1980), and succeeded in placing them on a preliminary map of the *Volvox* genome (Huskey et al, 1979). Currently, we are attempting to extend the pioneering studies of Starr and Huskey by saturating the genome with mutations that cause a breakdown of the germ-soma dichotomy, in hopes of eventually defining all of the major loci whose actions are required to establish the complete division of labor between somatic and reproductive cells that characterizes this species.

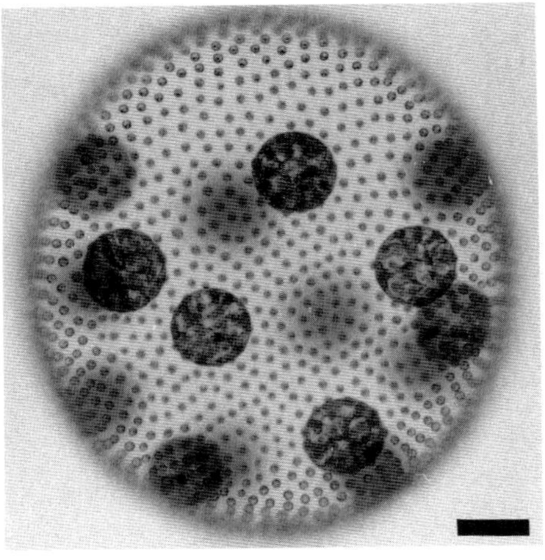

Figure 1. An asexual spheroid (individual) of *Volvox carteri* f. *nagariensis*. Each spheroid contains about 2000 small, *Chlamydomonas*-like, terminally differentiated somatic cells at the periphery of a transparent, glycoprotein-rich sphere, and about 16 large asexual reproductive cells, or "gonidia", in the interior. In each life cyle, the gonidia divide to produce juvenile spheroids that are miniatures of the adult. Bar: 50 μm.

---

[1]This paper is dedicated to Richard Starr on the occasion of his 65th birthday.

Normal patterns of germ-soma differentiation.

Dichotomous germ-soma development in asexual *V. carteri* f. *nagariensis* normally begins at the sixth embryonic cleavage division, when (optimally) the 16 cells in the anterior end of the embryo divide asymmetrically (Green and Kirk,1982) to produce large and small sister-cell pairs (Fig. 2). After dividing asymmetrically three more times (Green and Kirk, 1981), the larger cells (the reproductive, or "gonidial," initials) withdraw from the division cycle, while smaller cells (the somatic initials) continue to divide two or three more times. Following cleavage, the embryos turn inside out in a process called "inversion" (Viamontes et al, 1979). Only after inversion is completed, and only if the culture is illuminated, does functional differentiation of somatic cells and gonidia occur (Kirk and Kirk, 1985). During and following the period of cell differentiation, the spheroid is enlarged by deposition of a complex extracellular matrix (Kirk et al, 1986), in the absence of further cell divisions; about halfway through this expansion period, the juvenile individuals escape from the parent and swim away, whereupon the parental somatic cells become senescent and die (Fig 3). Although cleavage, inversion, cytodifferentiation and initial expansion normally take place inside a parental spheroid, they will occur identically in gonidia that have been experimentally released from the parent prior to the initiation of cleavage.

A sexual cycle can be induced in this organism by exposure of cultures to a powerful inducer (Starr and Jaenicke, 1974; Mages et al, 1988) that is produced by sexual males (Starr, 1970), or by heat-shocked asexual individuals (Kirk and Kirk, 1986). Following exposure to inducer, gonidia do not transform into gametes directly; instead, they initiate a new round of cleavage in which asymmetric division

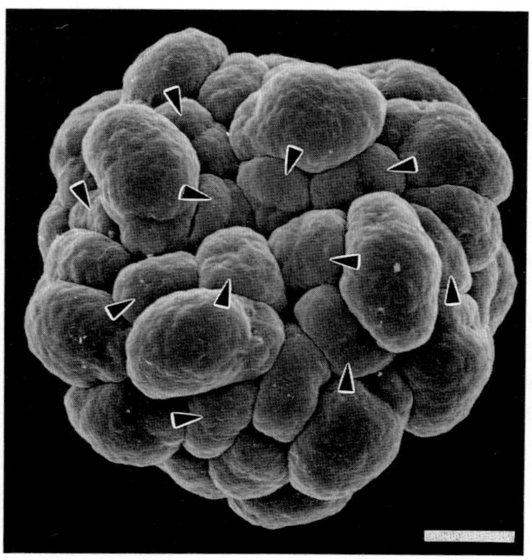

Figure 2. A scanning electron micrograph of a 64-cell asexual embryo of *V. carteri*, viewed from the anterior pole. Each arrowhead points from a gonidial-intial to the somatic-initial from which it has just been set apart by asymmetric division. An incipient furrow is visible toward the counterclockwise end of each gonidial initial, indicating where it was preparing to divide asymmetrically a second time, to generate another somatic initial. Bar: 10 µm.

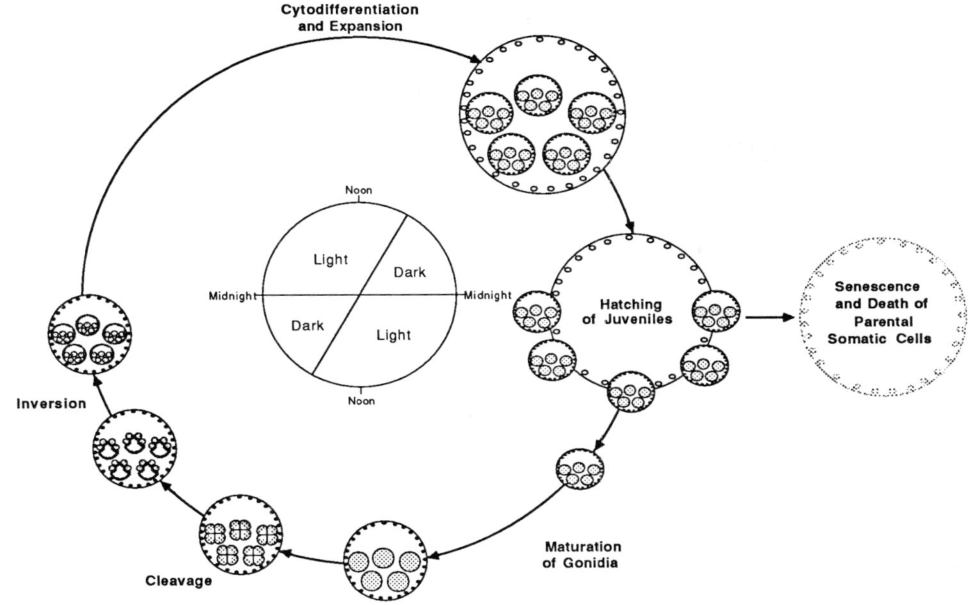

Figure 3. A diagram of the asexual life cycle of *V. carteri* f. *nagariensis* as synchronized by a light-dark cycle. Mature gonidia divide to produce embryos containing all of the cells that will be present in the adult. These embryos then turn inside-out, in the process of inversion, to assume the adult configuration. As soon as the culture is reilluminated after inversion, differentiation of the gonidial and somatic initials begins, and both the juvenile and the adult spheroids begin to expand by deposition of extracellular matrix. Midway through their expansion, juveniles are released from the parental spheroid, and a few hours later their gonidia initiate a new round of embryogenesis. Meanwhile, somatic cells of the parental spheroid approach senescence and death. (Reproduced, with permission, from Kirk, 1988.)

patterns are modified, and spheroids containing somatic cells and gametes are produced.

It is noteworthy that in *V. carteri*--in distinction to many other species of *Volvox*, as well as members of all other genera in the family Volvocaceae--cells destined to reproduce do not first develop vegetative structures such as functional flagella and eyespots; rather, they develop directly into reproductive cells (gonidia in the asexual phase, eggs in females, or sperm-producing androgonidia in sexual males; Starr, 1969). Nor do *V. carteri* cells that develop vegetative structures normally have any reproductive potential whatsoever; they are terminally differentiated somatic cells that are programmed for senescence and death. Thus the division of labor between somatic and reproductive cells is complete in *V. carteri*.

The role of asymmetric division in germ-soma differentiation.

In embryos entering the sexual cycle (Starr, 1969) and in various Mul (**mul**tiple gonidia) mutants (Huskey et al, 1979), both the location and the timing of asymmetric divisions differ from the pattern described above. For example, in embryos of sexually-induced wild-type females, and in asexual embryos carrying certain *mulB* alleles, asymmetric divisions occur in the seventh rather than the sixth cycle, and in as many as three-fourths of the blastomeres, resulting in up to 48 reproductive cells (Fig 4A). In sexual males, and in asexual embryos carrying *mulC* or *mulD* mutations (Huskey et al, 1979), all blastomeres cleave asymmetrically at the last (usually the eighth) division, producing a 1:1 ratio of reproductive and somatic cells. In contrast to these highly regular patterns, certain Mul mutants exhibit an essentially random pattern of asymmetric division and gonidial formation (Fig. 4B). But, whatever the pattern, or lack of pattern, in which asymmetric divisions occur in either the asexual or sexual cycle, in wild-type or mutant embryos, a simple relationship between cell size and fate is normally observed: larger cells produced by asymmetric divisions become reproductive initials, while their smaller sister cells become somatic initials. Thus, visibly asymmetric division appears to be tightly linked to specification of reproductive cells in this species.

Martin Pall (1975) took the preceding concept one step further and proposed that "cell size or something closely related to cell size is critical in determining reproductive vs. somatic cell differentiation" in *V. carteri*. Pall's proposal was based on his observations that because Pcd (**p**remature **c**essation of **d**ivision) mutant embryos ceased cleaving 5 to 7 divisions prematurely, they produced far fewer cells than normal--but of greatly increased average size--and that they went on to become adults with excessive numbers of reproductive cells, but few somatic cells. Although we were initially very skeptical of Pall's hypothesis (Kirk and Harper, 1986), a number of our recent observations on genetically and experimentally modified cleavage patterns are fully consistent with it (*vide infra*; plus Kirk and Kirk, to be published).

Genetic control of asymmetric division.

Although at least four *mul* loci (Huskey, et al 1979) appear to control the pattern and timing of asymmetric divisions, certain mutations of at least one locus abolish asymmetric divisions altogether. We call this locus *gls* (**go**nidia**les****s**), because mutant embryos that fail to execute asymmetric divisions fail to produce any gonidia at all (Fig. 4C). Obviously, failure to produce reproductive cells is a lethal trait; therefore, strong *gls⁻* alleles can only be recovered in the presence of a *regA* mutation (to be described below) that confers reproductive potential on the normally-sterile somatic cells. We have, however, recovered "quasi-gonidialess" (Q-Gls) mutants on a wild-type background; such mutants execute 0-4 asymmetric divisions and produce adults with 0-4 gonidia (Fig. 4D). These small numbers of gonidia are sufficient to maintain the strain. Recently, we have recovered partial revertants of *regA⁻,gls⁻* strains that also produce very small numbers of "true" gonidia (Fig. 4E).

In our original *gls⁻* strains, symmetric divisions occur perfectly normally, but no asymmetric divisions occur. Therefore, our working hypothesis is that *gls⁺* controls some component of a system required to modulate behavior of the *V. carteri* cytokinetic apparatus in order for division to occur in an asymmetric, rather than a symmetric, manner. We are now studying the cytoskeletal differences between symmetrically and asymmetrically dividing blastomeres of wild-type embryos, with the hope of eventually elucidating how the *gls* locus may act to permit division asymmetry. If (as we suspect but have not yet proved) partial reversions of the Gls phenotype usually involve mutations of second-site suppressors of *gls*, rather than secondary mutations at the *gls* locus itself, they may help to define additional components of the asymmetric division apparatus.

Genetic control of differentiation in large cells produced by asymmetric division.

The large cells produced by asymmetric division normally differ in their behavior from smaller sister cells in two ways: they withdraw from the cleavage cycle about three divisions sooner, and they differentiate directly into gonidia. But in the presence of a lesion at any one of a set of (at least four) loci which we call *lag* (**la**te **g**onidia), both of these behaviors are affected. In Lag embryos, presumptive gonidial initials produced by asymmetric division appear to continue dividing as long as the smaller somatic initials do; as a consequence, by the end of cleavage they are smaller than normal gonidial initials; yet they are larger than normal somatic initials (Fig. 4F). These cells of intermediate size then differentiate as larger-than-normal somatic cells (Fig. 4G). However, after about one day they dedifferentiate, enlarge, and redifferentiate as *bona fide* gonidia that go on to cleave and repeat the cycle (Fig. 4H).

The Lag phenotype resembles the normal pattern of reproductive cell development seen in many other species of *Volvox*, as well as in all of the colonial relatives of *Volvox* in the family Volvocaceae. Indeed, when Huskey's laboratory first observed mutants of this type, they used the epithet "aureus-like" for them

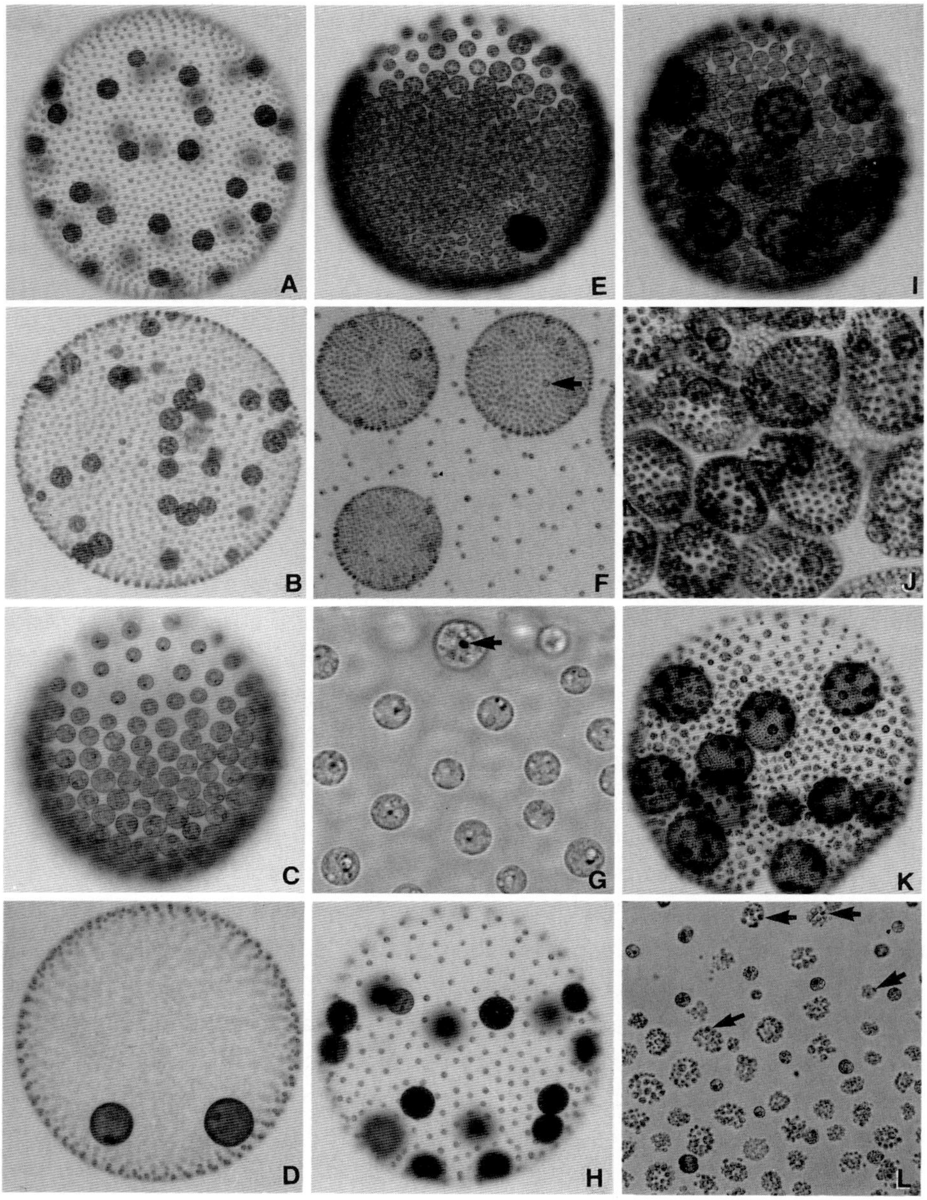

Figure 4. Selected mutant strains. **A, B**: "Multiple gonidia" (Mul) mutants. A: a *mulB* mutant with the asymmetric cleavage pattern, and reproductive cell distribution pattern, normally seen in sexual females. B: an unmapped Mul mutant exhibiting spatial and temporal variation in asymmetric cleavage divisions and variation in numbers, sizes and locations of gonidia. **C-E**: "Gonidialess" (Gls) mutants. C: a *gls, regA* mutant in which no asymmetric divisions occur, no "true" gonidia are formed, but (because of the *regA* mutation) somatic cells "regenerate" to form functional gonidia. D: a "quasi-gonidialess" mutant (Q-Gls, mutant locus

(Huskey, personal communication), to reflect the fact that the mutants resembled *V. aureus*, which is one of many *Volvox* species in which gonidia do not become visibly differentiated from somatic cells until after the young adults have hatched out of their parents and are free-swimming.

Because the phenotypes of strains mutant at any one or two of several unlinked *lag* loci (*lagA, lagB, lagC...*) are extremely similar, we assume that the *lag* loci encode products that act sequentially in a single pathway that is required to shortcut cleavage, bypass vegetative development, and shunt large cells directly into reproductive development.

Some insight into the possible role of cell size in activating the *lag* loci, and other genes involved in the program of gonidial development, may be derived from recent studies in which we have followed the development of cells of various sizes after using heat shock to interrupt cleavage of all blastomeres around the time of the eighth division (Kirk and Kirk, to be published). At this stage, somatic initials are presumably already too small to shut off division spontaneously when gonidial initials do (since they would normally continue dividing), but they are about eight times larger than somatic initials normally are at the end of cleavage, three cycle later. When cleavage of presumptive somatic initials is interrupted experimentally by the eighth division, many or all of these cells develop directly into gonidia. In contrast, if cleavage is stopped one cycle later, few of the presumptive somatic initials develop as gonidia.

The behavior of Lag embryos and heat-shocked wild-type embryos initially led us to the following extension of Martin Pall's hypothesis: There are two stages

---

not yet defined); the parental spheroid contained two gonidia, and each of these has cleaved to form a juvenile spheroid containing one gonidium. E: a partial revertant of the *gls,regA* mutant shown in C; one "true" gonidium was formed by asymmetric division. **F-H**: A "late gonidia" (Lag) mutant (mutation at *lagB*). F: juveniles within a parental spheroid have no obvious gonidia, but some slighly larger cells (arrow). G: higher magnification of one of the larger cells of a prehatching Lag juvenile; the prominent eyespot (arrow) and flagella (out of focus) indicate that this cell has differentiated as a somatic cell. H: about 36 hours post-hatching; normal-looking gonidia have developed and are about to cleave, but their development has been delayed by about one day relative to wild type. **I-L**: "Somatic regenerator" (Reg; *regA*) mutants contrasting the "standard Reg" phenotype (I,J), and the "modified Reg" phenotype (K,L) caused by a secondary mutation at the *ram* locus. In the standard Reg, regenerated somatic cells (I) and progeny spheroids derived from them (J) are large, fairly uniform in size, and mutually compressed by crowding. In the modified Reg, in contrast, some somatic cells do not ever divide, those that do divide enlarge little before division, and progeny "spheroids" are variable in size, shape and cell number (L). Arrows in L indicate large eyespots that persisted through regeneration and division and now reside in progeny cells that are scarcely larger than they are.

at which cell size plays a role in activating the program of gonidial differentiation, with different size thresholds operative in each case. First, cell size is "measured" after the eighth division and the *lag*+ loci are activated in cells that are above some threshold size, causing them to cease cleaving.[2] Then, size is "measured" again after the end of cleavage, and cells above some second size threshold (smaller than the first) are triggered to develop directly into gonidia.

The preceding hypothesis is unsatisfactory as an explanation of gonidial specification, however, in at least two regards. First, it is important to note that although *lag*- gonidial initials differ in their behavior from wild-type gonidial initials, they do eventually develop into gonidia. Second, it is important to note that although somatic initials of wild-type embryos are apparently too small to activate the *lag*+ loci (as evidenced by their failure to cease cleaving on their own initiative) they do develop as gonidia if their cleavage is interrupted experimentally. Therefore, although the *lag*+ loci appear to play an important role in establishing the complete division of labor between *V. carteri* somatic and reproductive cells, by controlling the timing and pathway of gonidial differentiation, they do not appear to be responsible for gonidial specification *per se*. If, as postulated by Pall (and as it currently seems to us) cell size is a causal factor in gonidial specification, it must act on some additional, as yet undefined genes that are more directly involved in the gonidial-specification process than the *lag* loci are. We consider identification of these genes an important challenge for the future.

Genetic control of differentiation in small cells produced by asymmetric division.

The smaller cells produced by asymmetric cleavage normally develop into terminally differentiated somatic cells that undergo programmed senescence and death (Pommerville and Kochert, 1981; 1982) and exhibit no reproductive potential whatsoever. But mutation at a single locus, the *regA* locus, confers on somatic cells all of the reproductive potentials of the species (Starr, 1970; Huskey and

---

[2] In principle, there are several ways in which quantitative differences in cell size might lead to qualitatively different cell behaviors. The following are just two examples: If all cells contained similar cytoplasmic concentrations of a sequestered gene activator that at some specific stage was released and became free to diffuse into the nucleus, only cells with a high cytoplasmic-to-nuclear ratio (and hence a high total amount of the activator) might experience activation of the corresponding gene. Alternatively, if all cells underwent a particular membrane-mediated triggering event (such a flux of calcium ions or protons) at some particular stage, small cells, with their lower cytoplasm-to-membrane ratio, might exhibit a qualitatively different response than larger cells, in which the cytoplasm would have much greater buffering capacity.

Griffin, 1979). Asexual Reg (somatic-**reg**enerator) spheroids look indistinguishable from wild-type spheroids initially; their somatic cells differentiate the eyespots, flagella and other features required for vegetative functions. But then they dedifferentiate, enlarge and redifferentiate as fully functional gonidia instead of undergoing programmed death (Fig. 4I, J). Somatic cells of sexually-induced Reg spheroids can also participate in gamete formation (Starr, 1970).

By now, more than 50 Reg mutants have been studied genetically and they all possess mutations that map to the *regA* locus. [The one strain that was described with a lesion at the *regB* locus (Huskey and Griffin, 1979) had a very different phenotype, in which cells with the normal features of somatic cells never developed, and hence there was no true somatic cell regeneration (Baran, 1984).]

Recently, we have discovered an extragenic suppressor of *regA*, which we call the *ram* (***regA*** **m**odifier) locus (Kirk, Stamer and Casteel, unpublished). In mutants with the $regA^{-}ram^{-}$ genotype (and hence the "modified-" or "M-Reg" phenotype), the asexual reproductive behavior of somatic cells differs from that of St-Reg ("standard regenerator") mutants in at least four regards (Fig. 4K, L): (i) Somatic cells enlarge less and produce more extracellular matrix prior to division; therefore, M-Reg spheroids are larger and much less dense appearing than St-Reg spheroids. (ii) Regeneration and division of somatic cells is less uniform; therefore, cells at many different stages of regeneration are often seen within a single region, and some somatic cells never divide. (iii) Regenerating somatic cells begin dividing at much smaller average sizes; therefore, much smaller progeny spheroids, sometimes consisting of only a few somatic cells, are produced. (iv) Dedifferentiation of somatic cells (*e.g.* resorption of flagella and eyespots) prior to division is often incomplete; therefore, M-Reg spheroids retain motility throughout regeneration (whereas St-Reg spheroids do not), and they frequently produce progeny in which the eyespot of the parental somatic cell has been carried through division and occupies more than half the volume of the progeny somatic cell that inherits it. All of these traits indicate a partial suppression of reproductive potential in $regA^{-}$ somatic cells when a ram^{-} allele is also present.

Crosses of M-Reg strains to wild type produce St-Reg, M-Reg, wild-type, and inviable progeny in approximately a 1:1:1:1 ratio[3] (Stamer and Kirk, unpublished). Test cross results are consistent with the conclusion that the genotypes of the three surviving categories of F1 progeny are $regA^{-}ram^{+}$, $regA^{-}ram^{-}$, and $regA^{+}ram^{+}$, respectively. This implies that the inviable progeny are $regA^{+}ram^{-}$, suggesting that

---

[3] Note that since *Volvox* is haploid in all active stages of the life cycle, there are no dominant or recessive traits; thus, in a cross between parents differing at two independently segregating loci, four progeny classes in a 1:1:1:1 ratio are to be expected.

the *ram⁻* allele is lethal in the absence of a *regA⁻* allele. Because the *ram⁻* allele suppresses aspects of the reproductive potential of Reg somatic cells, we presume that its lethality on a wild-type background may be due to a suppression of reproductive potential in "true" gonidia. Consistent with this suggestion is our observation that the "true" gonidia of *regA⁻ram⁻* strains often possess eyespots, whereas the gonidia of wild-type or standard Reg strains never do.

Because mutation at the *regA* locus confers the entire spectrum of asexual and sexual reproductive potentials on cells that are normally sterile, we have postulated that *regA⁺* encodes a negative regulator of reproductive development (Kirk and Harper, 1986). Following parallel logic, the fact that *ram⁻* suppresses the *regA⁻* phenotype in double mutants suggests that *ram⁺* is a positive regulator of reproductive development. An interesting possibility, worthy of future study, is that *regA* encodes a transcriptional inhibitor, and *ram* encodes a transcriptional activator, for a whole battery of genes required for reproductive development.

<u>Regulation of *regA* expression.</u>

If *regA* encodes a negative regulator of reproductive development, it follows that the locus must be fully expressed in wild-type somatic cells (which are sterile under all conditions that have been tested), but completely inactive in gonidia (which can be prevented from reproducing only by withholding energy or poisoning them). How might such "fully-on/fully-off" regulation of *regA* be achieved?

Studies of its unusual mutational properties raised the intriguing possibility that the *regA* locus may be turned off and on by cyclic DNA rearrangement (Kirk et al, 1987). The *regA* locus becomes hypermutable during brief periods just before and just after each cycle of embryogenesis. During these two periods, exposure to agents that induce error-prone recombination and repair--but not exposure to conventional point mutagens--results in a *regA* mutation rate as high as ten percent (which is nearly 10^5 times the spontaneous rate). This extraordinary, locus-specific hypermutability occurs in the gonidia, where the gene is never expressed, not in the somatic cells where the gene is expressed. And, the period of hypermutability just after embryogenesis appears to precede by only a few hours the time when temperature-sensitive-mutant studies suggest that the *regA* locus normally acts in somatic cells (Huskey and Griffin, 1979). These results are consistent with the following three-part hypothesis: (i) the *regA* locus exists in two configurations, one of which permits, and one of which precludes, its activity; (ii) the locus switches from the "on" to the "off" configuration in gonidia at the end of embryogensis (permitting these cells to reproduce), and then switches back to the "on" configuration just prior to cleavage (so that it will be inherited by all progeny cells in that state); and (iii) if error-prone recombination has been induced prior to one of these switching periods, deletion of the rearranging portion frequently occurs. One possible version of this hypothesis is diagrammed in Fig. 5.

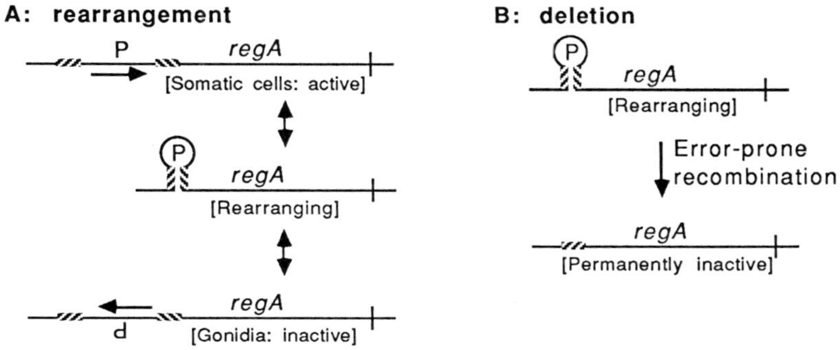

Figure 5. A model to explain the regulation and hypermutability of the *regA* locus. **A**: It is proposed that regulation of expression occurs, as in the H2 gene of *Salmonella* (Zeig et al, 1978), by reversible inversion of the promoter region (P) lying between two inverse-repeat elements (hatched bars). **B**: If the foregoing were the case, the presence of an induced error-prone recombination system in the cells at the time of rearrangement could result in frequent deletion of the rearranging region.

While all indirect tests performed to date are consistent with the preceding hypothesis, the only definitive test will be to determine whether the gene differs in its sequence in gonidia *vs.* somatic cells. This, along with a desire to identify the *regA* product and study its mechanism of action, motivates our present attempts to clone the locus.

<u>Attempts to clone the *regA* locus.</u>

Because nothing is yet known about the molecular nature of the *regA* product, an indirect approach to cloning was required. Therefore, we initiated mapping of the *Volvox* genome by use of restriction fragment length polymorphisms (RFLPs), in search of an identifiable DNA fragment that is tightly linked to the *regA* locus and could be used as the starting place for a "chromosome walk" to *regA*. Using middle-repetitive DNA elements as probes, we identifed two alleleic pairs of *regA*-linked RFLPs (Harper et al, 1987). Because the probes used to detect these linked polymorphic elements each identify numerous sequences throughout the genome, the walk could not begin until a unique portion of the linked polymorphic region had been cloned. This has been accomplished (Fig. 6) and a chromosome walk from the cloned element to the *regA* locus is now in progress (Miller and Kirk, to be published).

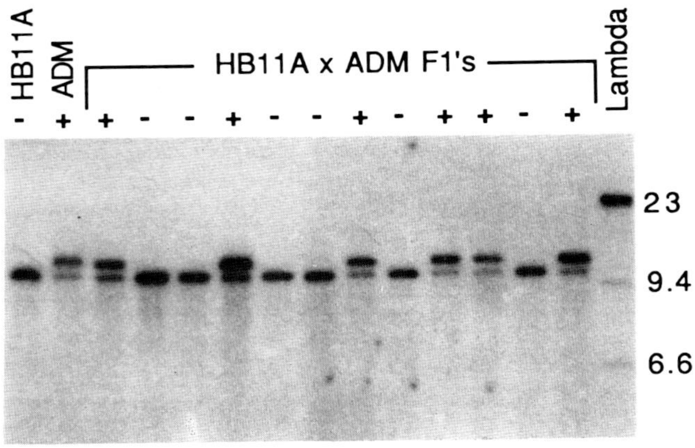

Figure 6. A Southern blot of genomic DNA from a *regA*⁻ female (HB11A), a wild-type male (Adm), and 12 of their progeny. DNA samples were digested with *Sau*3A and the blot was probed with a cloned genomic fragment called "2-8-4." The *regA* genotype of each of the progeny is given above the coresponding lane (+ = wild type and - = mutant). In the progeny sample shown here, alleles at the the *regA* locus cosegregated perfectly with the allelic DNA fragments detected by 2-8-4 hybridization. However, two recombinant progeny were detected in the set of 89 analyzed, indicating that the polymorphic DNA detected by 2-8-4 lies about two centimorgans from the *regA* locus.

Conclusion.

The promise that Weismann saw in *Volvox* 100 years ago as a model for analyzing the mechanism of dichotomous differentiation of germ from soma is beginning to be realized. The major breakthroughs all occurred in Richard Starr's laboratory, where *Volvox* was first brought under continuous cultivation (Darden, 1966), the life cycle of *V. carteri* was first described in detail (Kochert, 1967), and *V. carteri* f. *nagariensis* was first shown to have the attributes necessary for genetic analysis of development (Starr, 1970). Huskey and his colleagues moved the problem forward substantially by revealing more of the developmental-genetic potentials of this forma, and by developing a preliminary map of its genome (Huskey et al, 1979). Now the organism appears to be indicating (albeit somewhat reluctantly at times) its willingness to share some of its secrets at the molecular level. Furthermore, the existence of a number of simpler--but apparently closely related--unicellular and colonial relatives of *Volvox* (Kochert, 1975) holds out hope that once the molecular-genetic basis for the germ-soma division of labor in *V. carteri* comes to be better understood, a molecular understanding of how such a developmental-genetic program evolved may be accessible. It is to be hoped that much of the potential that we now perceive in this system will be realized long before another century has passed.

Acknowledgements. Appreciation is expressed to Kandace A. Stamer for performing the tedious and difficult unpublished Mendelian-genetic studies reported here, and to Christine R. Adams, J. Allen Casteel, Marilyn M. Kirk and Judith K. Miller for performing various other unpublished studies that are discussed. This research was supported by grant No. GM-27215 from the National Institutes of Health and grant No. DMB-8615691 from the National Science Foundation.

References.

Baran G (1984) Analysis of somatic cell differentiation in *Volvox carteri* f. *nagariensis*. PhD. thesis, University of Virginia, Charlottesville

Callahan AM, Huskey RJ (1980) Genetic control of sexual development in *Volvox*. Develop Biol 80:419-435.

Darden WH (1966) Sexual differentiation in *Volvox aureus*. J Protozool 13:239-255

Green KJ, Kirk DL (1981) Cleavage patterns, cell lineages, and development of a cytoplasmic bridge system in *Volvox* embryos. J Cell Biol 91:743-755

Green KJ, Kirk DL (1982) A revision of the cell lineages recently reported for *Volvox carteri* embryos. J Cell Biol 94:741-742

Harper JF, Huson KS, Kirk DL (1987) Use of repetitive sequences to identify DNA polymorphisms linked to *regA*, a developmentally important locus in *Volvox*. Genes & Develop 1:573-584

Huskey RJ, Griffin BE (1979) Genetic control of somatic cell differentiation in *Volvox*: analysis of somatic regenerator mutants. Develop Biol 72: 226-235

Huskey RJ, Griffin BE, Cecil PO, Callahan AM (1979) A preliminary genetic investigation of *Volvox carteri*. Genet 91:229-244

Kirk DL (1988) The ontogeny and phylogeny of cellular differetniation in *Volvox*. Trends Genet 4:32-36

Kirk DL, Birchem R, King N (1986) The extracellular matrix of *Volvox*: a comparative study and proposed system of nomenclature. J Cell Sci 80:207-231

Kirk DL, Harper JF (1986) Genetic, biochemical and molecular approaches to *Volvox* development and evolution. Int Rev Cytol 99:217-293

Kirk MM, Kirk DL (1985) Translational regulation of protein synthesis, in response to light, at a critical stage of *Volvox* development. Cell 41:419-428

Kirk DL, Kirk MM (1986) Heat shock elicits production of sexual inducer in *Volvox*. Science 231:51-54

Kirk DL, Baran GJ, Harper JF, Huskey RJ, Huson KS, Zagris N (1987) Stage-specific hypermutability of the *regA* locus of Volvox, a gene regualting the germ-soma dichotomy. Cell 48:11-24

Kochert G (1967) Differentiation of reproductive cells in *Volvox carteri*. J Protozool 15:438-452

Kochert G (1975) Colony differentiation in green algae. In: Coward SJ (ed) Developmental regulation: aspects of cell differentiation. Academic, New York

Mages H-W, Tschochner H, Sumper M (1988) The sexual inducer of *Volvox carteri*: primary structure deduced from cDNA sequence. FEBS Lett 234:407-410

Pall ML (1975) Mutants of Volvox showing premature cessation of division: evidence for a relationship between cell size and reproductive cell differentiation. In: McMahon D, Fox CF (eds) Developmental biology: pattern formation, gene regulation. Benjamin, Menlo Park

Pommerville JC, Kochert GD (1981) Changes in somatic cell structure during senescence of *Volvox carteri*. Eur J Cell Biol 24:236-243

Pommerville J, Kochert G (1982) Effects of senescence on somatic cell physiology in the green alga *Volvox carteri*. Exp Cell Res 140:39-45

Sessoms AH, Huskey RJ (1973) Genetic control of development in *Volvox*: isolation and characterization of morphogenetic mutants. Proc Natl Acad Sci USA 70:1335-1338

Starr RC (1969) Structure, reproduction and differentiation in *Volvox carteri* f. *nagariensis* Iyengar, strains HK 9 and 10. Arch Protistenkd 111:204-222

Starr RC (1970) Control of differentiation in *Volvox*. Dev Biol (suppl) 4:59-100

Starr RC, Jaenicke L (1974) Purification and characterization of the hormone initiating sexual morphogenesis in *Volvox carteri* f. *Nagariensis* Iyengar. Proc Natl Acad Sci USA 71:1050-1054

Viamontes GI, Fochtmann LJ, Kirk DL (1979) Morphogenesis in *Volvox*: analysis of critical variables. Cell 17:537-550

Vig BK, Lewis R (1978) Genetic toxicology of bleomycin. Mutation Res 55:121-145

Weismann A (1889) The continuity of the germ plasm as a foundation of a theory of heredity. In: Poulton EB, Schonland S, Shipley AE (eds) Essays upon heredity and kindred biological problems. Clarendon, Oxford

Zeig J, Hilmen M, Simon M (1978) Regulation of gene expression by site-specific inversion. Cell 15:237-244.

Evolution of cytokinesis in green algae: A reticulum of tubular membranes characterizes septum membrane biogenesis in Ulvophyceae

H.J. Sluiman[1] and P.C.J. Blommers
Rijksherbarium and Department of Plant Molecular Biology
University of Leiden
Postbus 9514
NL-2300 RA Leiden
The Netherlands

Introduction

Ideas about green algal evolution have undergone considerable changes during the last decades, yet certain basic assumptions dating back from the beginning of this century remained unchallenged. One of these assumptions still widely accepted is that multicellular plants, including green algae, stem from unicellular ancestral flagellates. Presumably, from this basic level of organization more advanced forms evolved step by step (see discussion and references in Sluiman 1989, Sluiman *et al.* 1989). Recent studies using ultrastructural and other modern analysing techniques, however, have provided evidence that in green algae at least three separate evolutionary lineages may be envisaged, with each lineage having its own distinct origin among the hypothetical assemblage of ancestral flagellates. Members of each lineage are identifiable as such because they have certain flagellar apparatus features in common, features that they are supposed to have inherited from their flagellated predecessors. Members of a lineage, however, may differ in other respects, such as, e.g., structure and organization of the plant body, and the

---

[1] Present address: Universität zu Köln, Botanisches Institut, Gyrhofstraße 15, D-5000 Köln 41, FRG

mechanism of cellular division. One of the general goals of the present study is to reconstruct the possible evolution of cytokinesis that must have occurred in each lineage.

In previous reviews dealing with cytokinetic diversity in green algae, division models (categories) were described only in very general terms (e.g., Mattox and Stewart 1984). For example, cellular division was often described as being effected either by "furrowing" of the plasma membrane (PM), or by "cell plate formation". As subsequent studies have shown, however, cellular division is a structurally and dynamically highly complex phenomenon that must be characterized more precisely. Cytokinetic features requiring special attention include (a) septum PM biogenesis (membrane flow); (b) pattern of cell wall deposition; (c) behaviour and involvement of the cytoskeleton (e.g., the phycoplast); (d) presence and behaviour of flagellar apparatus components (basal body apparatus) (see Sluiman *et al*. 1989 for references and discussion).

It is unfortunate that progress in elucidating the ultrastructural details of cytokinesis in the green algae has been relatively slow in recent years, especially when compared to the advances made in flagellar apparatus research. Some progress towards understanding the phylogeny of cytokinesis is currently being made in the chlorophycean lineage. Basically, two major classes of division may be distinguished, "*Sporulation*" and "*Zellteilung*", on basis of careful light microscopical and electron microscopical examination (Ettl 1988, Sluiman *et al*. 1989). According to this preliminary concept, Sporulation is characterized in general (1) by the division product being a zoid (e.g., zoospore) or a modified (or reduced) zoid (e.g., autospore), (2) by a persisting flagellar apparatus that may be reduced to varying degrees (e.g., as a "basal apparatus") yet shows a characteristic behaviour and involvement with the cytokinetic process, (3) by a membrane flow pattern involving an ER → PM pathway (ER-plate *sensu* Sluiman 1989:86) or a Golgi → PM pathway (Kouwets, Blommers and Sluiman, unpublished obs.), and (4) by new cell walls (if produced) being deposited as a structurally distinct layer all around each daughter cell. In Zellteilung (= vegetative division, or desmoschisis), on the other hand, the progeny is *not* a zoid or a modified zoid, the flagellar apparatus (of which only the basal bodies have been retained) is *not* involved with septum formation, the septum PM is of Golgi origin ("cell plate"), and initial wall deposition is largely restricted to the new septum. The more primitive Chlorophyceae (Volvocales, Chlorococcales) are only able to undergo Sporulation-type divisions, whereas in the more advanced representatives (including Chaetophorales) both kinds of division may occur in a given species, serving distinct biological purposes (growth *versus* reproduction). The question which division categories occur in the Oedogoniales cannot be resolved on basis of presently available information.

Cytokinesis in ulvophycean algae

In the ulvophycean lineage (*sensu* Sluiman 1989) the lack of knowledge of the ultrastructural details of cytokinesis has so far frustrated any attempt to develop a concept of cytokinetic evolution. Therefore, it is not yet certain whether the concepts of Sporulation and Zellteilung (see above) will be applicable to Ulvophyceae in much the same way as in Chlorophyceae. Cellular division in various representatives has been studied (for references see Sluiman 1989), but in nearly all cases the published micrographs and descriptions leave essential questions about endomembrane flow and cytoskeletal involvement in the cytokinetic process unanswered. For example, septum formation in cellular (i.e., non-siphonous) Ulvophyceae has been described as being "initiated by centripetal ingrowth of the PM" (Sluiman 1989:85), but how membrane is incorporated and where it originates is mostly unknown. Also, the involvement of the microtubular cytoskeleton in the division process (which may now be considered to be a ubiquitous phenomenon in plant cells) has been suggested in a few cases ("*Cylindrocapsa involuta*", *Cladophora*, *Urospora*, *Ulothrix*, *Pleurastrum*, for references see Sluiman 1989), but little if anything is known about the development and spatial distribution of these microtubular systems.

The main reason for this lack of information seems to be the quality of ultrastructural preservation that has been obtained using "traditional" fixation protocols (i.e., using aqueous aldehydes and OsO_4). In a previous study, we were able to obtain an improved resolution of cytoplasmic detail in dividing cells of the coccoid ulvophycean alga, *Trebouxia aggregata* by employing freeze fixation/freeze substitution techniques (Sluiman and Lokhorst 1988). We observed excellent preservation of organelles such as microtubules, the Golgi apparatus, and a partially coated *trans*-Golgi reticulum (PCR). In view of these results, it was decided to investigate cellular division in a second representative of the Ulvophyceae, the weakly filamentous *Gloeotilopsis planctonica*, using freeze fixation and freeze substitution.

Gloeotilopsis planctonica Iyengar *et* Philipose (strain CS 31-1, isolated from soil, deposited at the Sammlung von Algenkulturen in Göttingen, FRG) was prepared for electron microscopy essentially as described in Sluiman and Blommers (1989). The organism belongs to the Ulvophyceae because of the following features: in the flagellar apparatus the two basal bodies show a counterclockwise rotational displacement (when viewed from the anterior) and are overlapping, the zoospores are scale-covered, and the basal apparatus (BA) position is distinctly lateral to the mitotic spindle poles (present paper and unpubl. obs.). In this report we shall focus our attention to the ultrastructural aspects of membrane flow during cellular division.

A *trans*-Golgi reticulum is involved with cytokinetic septum membrane biogenesis in *G. planctonica*

In interphase cells of *G. planctonica* the nucleus occupies a slightly eccentric position, and is associated on one side with the most prominent organelle: the plate-shaped, parietal chloroplast. The opposite, "cytoplasmic" side is invariably marked by the presence of the persisting pair of basal bodies (basal

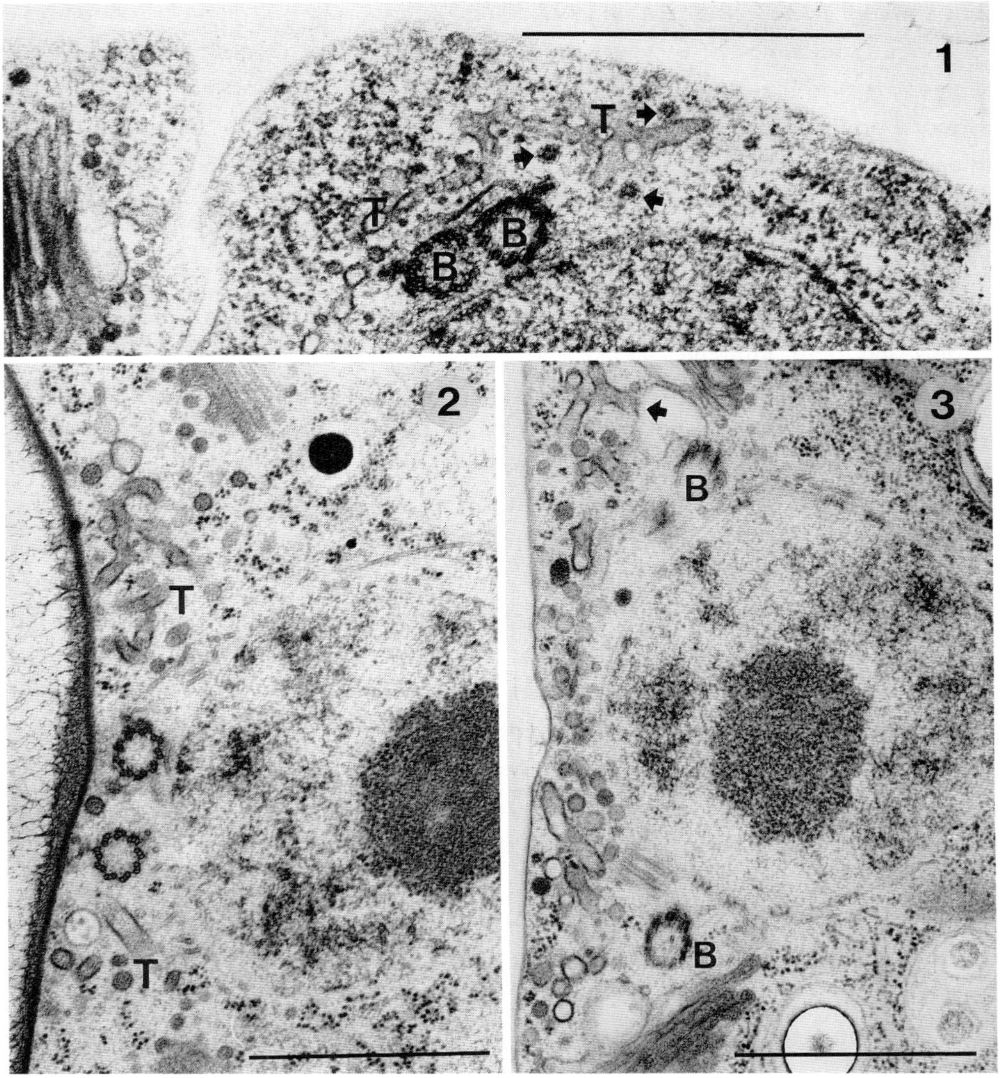

Figs. 1-3. *G. planctonica*. Scale bars = 1 µm. Fig. 1. Interphase cell with basal apparatus (B) near nuclear envelope, and partially coated (arrows) *trans*-Golgi reticulum (T). Fig. 2. Cell shortly before prophase. Fig. 3. Slightly later stage, with separated daughter basal apparatuses. Possible TGR-Golgi continuity marked by arrow.

apparatus: BA). The BA surrounding cytoplasm is rich in endomembranes, including a system of branched tubular membranes showing some similarity to the "Partially Coated Reticulum" (PCR) as described by Pesacreta and Lucas (1984, 1985) (fig. 1). The diameter of the tubules is in the order of 60-90 nm. The coating layer proved more difficult to preserve than in previous studies on green algae (e.g., Sluiman and Lokhorst 1988).

As the cell prepares for division, the BA and the membrane reticulum (hereafter: *trans*-Golgi reticulum, or TGR) are located close to the plasma membrane (fig. 2). The BA duplicates, and daughter-pairs of basal bodies move into opposite directions along the nuclear surface into the direction of the forthcoming spindle poles (fig. 3). [Like in other Ulvophyceae (see Sluiman 1989), the BA's stop migrating when they are about 1-2 basal body diameter away from the site to which the spindle microtubules will be focussed.] The TGR now expands progressively and completely fills the perinuclear, cortical cytoplasm, while forming a structural continuum with the inflated, *trans*-Golgi cisternae (figs. 3,4). The latter are normally devoid of any recognizable contents, but in germinating zoospores (ready to undergo their first division) they contain particles identical to the scales present on the zoospore and on the outer cell walls of young germlings (not shown). Conclusive evidence of membrane continuity between the *trans*-Golgi scale cisternae and the TGR was not obtained, but in various sections an intimate physical contact between the two endomembrane compartments was observed (figs. 3-6, 8).

During all mitotic stages the TGR is the most prominent organelle in the perinuclear cortical cytoplasm, and remains closely associated with the *trans*-Golgi cisternae. At telophase further TGR proliferation occurs into the interzonal (internuclear) cytoplasm (fig. 6). This coincides with the appearance of microtubules (MTs) near the TGR. Our serial section analysis has indicated that this MT system is part of a larger, apparently BA-organized array that also includes nuclear envelope-associated MTs (some of these nucleus-ensheathing MTs are visible in fig. 6). The BA-organized MT system appears to be structurally distinct from the MTs representing the remnants of the interzonal mitotic spindle.

The TGR plays a key role in the formation of the new septum at cytokinesis. Septum development is initiated as a unilateral invagination of the plasma membrane (PM) into the internuclear cytoplasm. The advancing furrow is closely associated with elements of the TGR, in micrographs often appearing as vesicle-like profiles, suggesting that TGR membranes are incorporated into the cleavage furrow. Some, necessarily "static", pictorial evidence of membrane fusion just having occurred is presented in fig. 7. This section was taken roughly in a plane perpendicular to the furrow advancement axis, and cuts through the leading (anterior-most) part of the furrow. It shows that MTs are involved with the process of TGR → PM transfer (in fact, exocytosis). These MTs are part of the BA-organized system mentioned above. Profiles of membranes other than the TGR are not observed in the area of cytokinesis. The spatial relationships of the Golgi apparatus, scale-cisternae, TGR, and furrow PM during cytokinesis are clearly illustrated in fig. 8, which shows the advancing septum in "longitudinal" section.

The TGR: a typical constituent of the ulvophycean membrane flow pathway at cytokinesis?

Our observations on cytokinesis in *G. planctonica* suggest a key role of the TGR in the ontogeny of the newly formed septum that separates daughter cells. The TGR does not appear very active in interphase cells (considering its relatively small size) but expands and proliferates considerably in mitotic cells. After telophase it forms a dense network of tubular membranes spanning the *trans*-Golgi area (near the former spindle poles) and the cleavage septum. Thus, it seems likely that the TGR is part of an over-all transport route of membranes running from the endoplasmic reticulum, via the Golgi apparatus and the TGR, to the septum PM. Circumstantial evidence in support of the specific role of the TGR in the septum biogenetic process includes (1) the

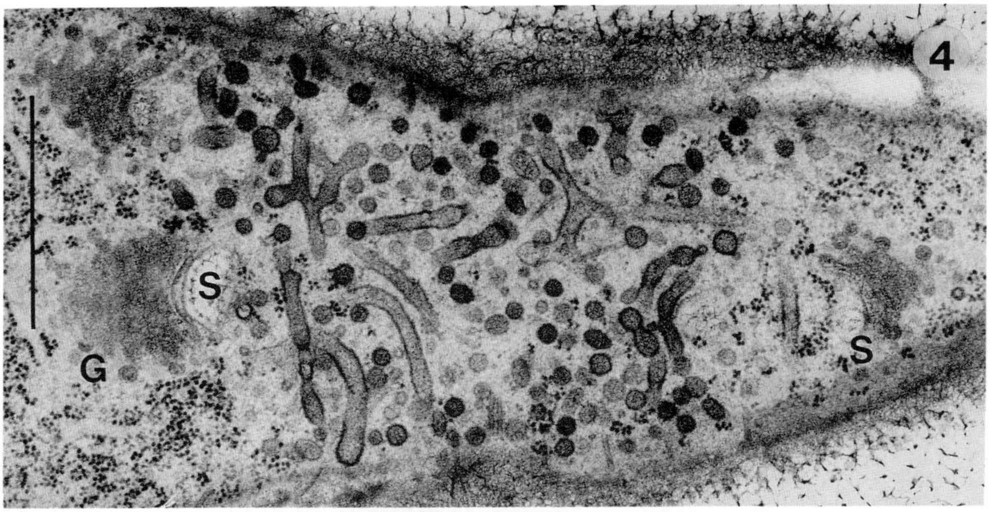
4
S
G
S

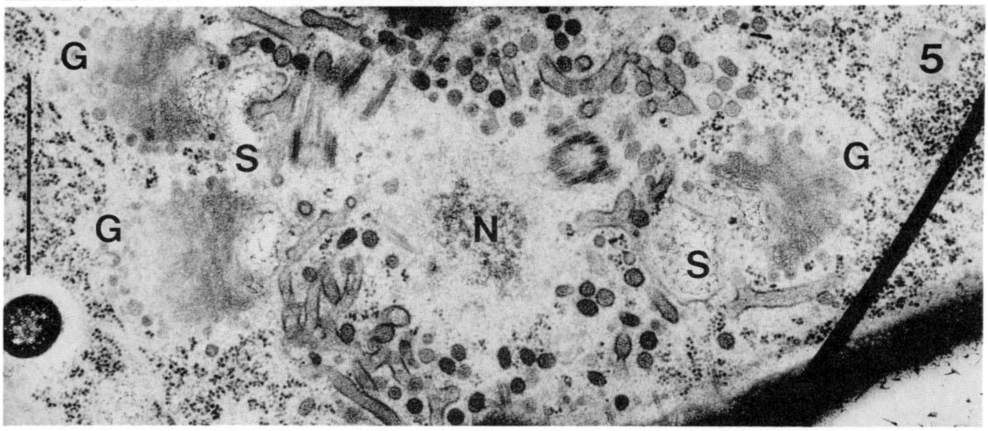
G
5
S
G
G
N
S

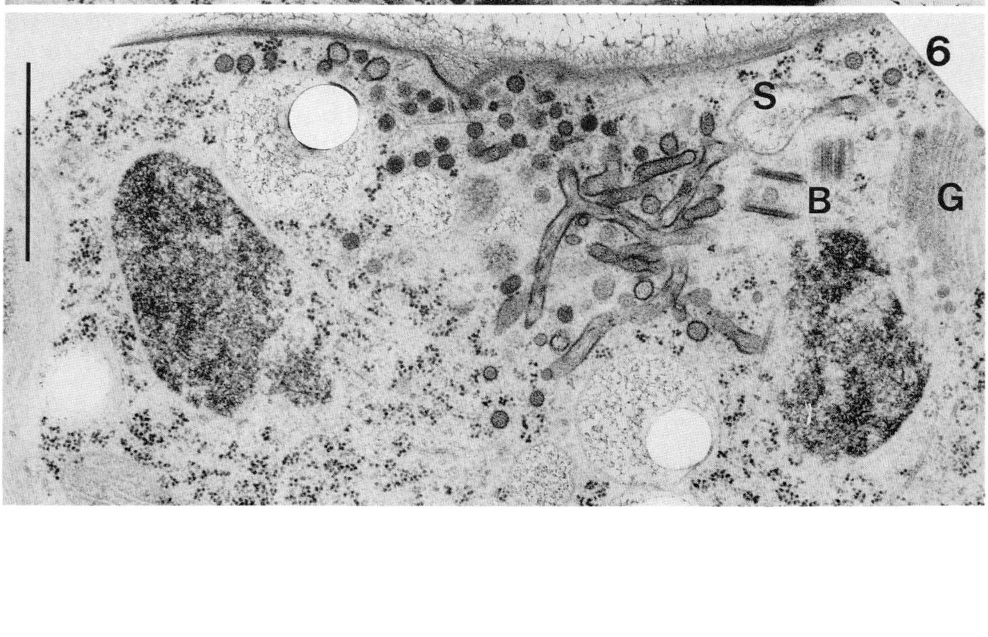
6
S
B
G

Figs. 4–6. *G. planctonica*. Scale bars = 1 µm. Figs. 4, 5. Same stage as fig. 3. Sections nrs. 6 and 8 from a series cut in a plane parallel to the mitotic axis, through the cortical, TGR-containing (T) cytoplasm. Notice scale containing cisternae (S) at *trans*-side of Golgi stacks (G), closely associated (continuous?) with TGR tubules. N, nucleus. Fig. 6. Telophase. The TGR, closely associated with scale cisternae (S), invades the interzonal cytoplasm. B, basal apparatus.

fact that the TGR is reduced (in size) to a minimum in non-dividing cells; (2) the observation that TGR proliferation is restricted both temporally and spatially, i.e. it occurs only when cytokinesis is imminent, and is only observed in areas of PM growth, not in other cortical parts of the dividing cell.

A similar cytokinetic involvement of branched tubular membranes, associated with MTs, was observed in *Trebouxia aggregata* by Sluiman and Lokhorst (1988). They noted that this type of membrane flow at cytokinesis was previously not known in green algae. One reason seems to be that all previous studies on cell division utilized fixation methods based on aqueous solutions of aldehydes and osmium, whereas Sluiman and Lokhorst were the first to employ freeze fixation and freeze substitution techniques in a cell division study which generally give a superior ultrastructural preservation (see also Lokhorst, Sluiman and Star 1988.) Tubular membrane reticula may be too sensitive for, e.g., osmotic shocks to be adequately preserved following "traditional", routinely applied aqueous fixatives. Indeed, in *G. planctonica* cells, sequentially fixed in buffered 0.6 or 1.7 % glutaraldehyde and in 1% OsO_4, we found a disorganized mass of vesicles instead of the reticulum preserved after freeze fixation (unpubl. obs.)

We believe that the intercalation of a *trans*-Golgi reticulum of tubular membranes in the PM biogenetic pathway at cytokinesis may be characteristic of members of the Ulvophyceae (*sensu* Sluiman 1989). Thus, this phenomenon has the potential of becoming another distinctive feature (apart from flagellar apparatus characters) of this evolutionary lineage. In the only Ulvophyceae that have been freeze-fixed so far, *Trebouxia* (Sluiman and Lokhorst 1988) and *Gloeotilopsis* (present study),

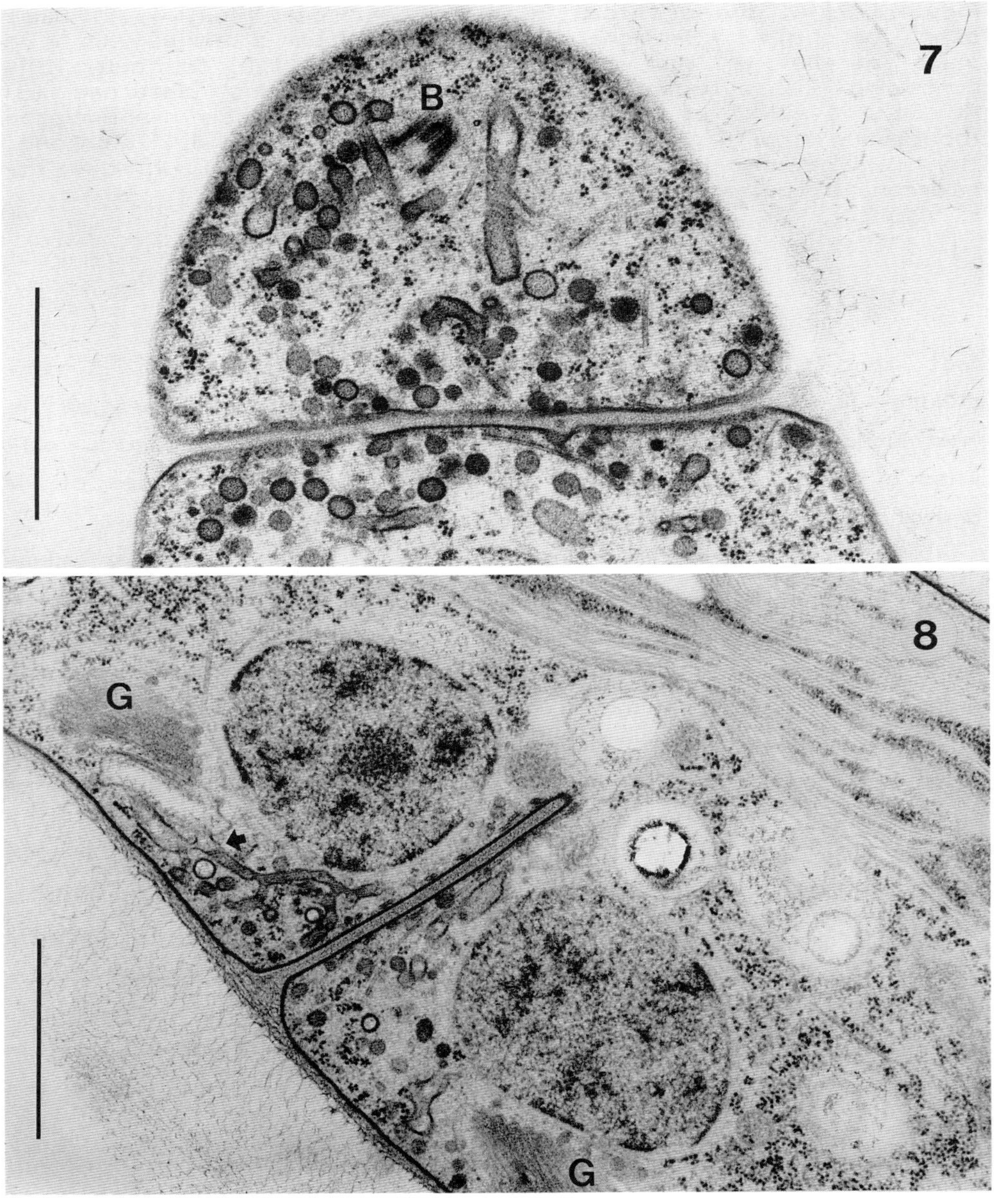

Figs. 7, 8. *G. planctonica*. Scale bars = 1 µm. Advancing cleavage furrow cut in two different planes (see text) after completion of mitosis. The septum is closely associated with TGR profiles and microtubules. Notice basal apparatus (B) position at a distance from the septum (fig. 7). Possible continuity of TGR and scale cisterna visible in fig. 8 (arrow in fig. 8). G, Golgi.

it was established that TGR-mediated cytokinesis occurs. Preliminary results by P.J. Segaar (pers. comm.) indicate that this also applies to *Friedmannia israelensis* (which also belongs to the Ulvophyceae *sensu* Sluiman 1989). All other ultrastructural studies of cell division in ulvophycean algae have utilized aqueous fixatives. Interestingly, in essentally all published micrographs the cytoplasmic area between the mitotic nucleus and the furrow initiation site (where in *G. planctonica* and in *T. aggregata* an extensive TGR is found), a disorganized collection of variously sized and shaped vesicles is visible. Also, this vesiculate cytoplasm is predominantly located at the *trans*-sides of Golgi stacks (see various published micrographs of, e.g., species of *Trichosarcina*, *Pseudendoclonium*, *Ulothrix*, *Ulva*, "*Cylindrocapsa*", *Urospora*, for references see Sluiman 1989).

In the other two major evolutionary groups of green algae, the classes Chlorophyceae and Charophyceae, there is no evidence for a TGR-mediated cytokinetic mechanism. In these groups, most ultrastructural studies using aldehyde fixatives have made reasonably clear that the septum membrane is either ER-derived or Golgi-derived. In those cases where the details of septum formation were not elucidated, there appears to be no reason so far to suspect a TGR-involvement because nondescript, disorganized vesicular aggregates which could be the artefactual remnants of a membrane reticulum are not observed.

Recently advanced hypotheses on green algal evolution stipulate that the classes Ulvophyceae, Chlorophyceae, and Charophyceae represent distinct phyletic lines that go back to different ancestral green flagellates (Mattox and Stewart 1984). This implies that features pertaining to the vegetative plant body such as, e.g., "true" cell walls, and "vegetative cell division" (desmoschisis, or Zellteilung sensu Ettl 1988), which do not occur in flagellates, evolved independently in each line. If this view is correct, it would not be surprising to find that cells of Ulvophyceae and other green algae have developed different methods (especially at subcellular level) to undergo vegetative division (Zellteilung sensu Ettl). It is hypothesized that one of the differences could be the involvement of a TGR in septum formation in Ulvophyceae.

Acknowledgements

We thank Dr J. Komárek, Czechoslovak Academy of Sciences, Třeboň, ČSSR, for his expert guidance during a field excursion near Třeboň from which the isolation of *Gloeotilopsis planctonica* (strain CS 31-1) eventually resulted. HJS is supported as a C. and C. Huygens-fellow by the Netherlands Organization for Scientific Research (NWO), subsidy nr. H 86-88.

References

Ettl H (1988) Zellteilung und Sporulation als wichtige Unterscheidungsmerkmale bei Grünalgen (Chlorophyta). Arch Protistenkd 135:103-118

Lokhorst GM, Sluiman HJ, Star W (1988) The ultrastructure of mitosis and cytokinesis in the sarcinoid *Chlorokybus atmophyticus* (Chlorophyta, Charophyceae) revealed by rapid freeze fixation and freeze substitution. J Phycol 24:237-248

Mattox KR, Stewart KD (1984) Classification of the green algae: a concept based on comparative cytology. In: Irvine DEG, John DM (eds) Systematics of the green algae. Academic Press, London Orlando, p 29

Pesacreta TC, Lucas WJ (1984) Plasma membrane coat and a coated vesicle-associated reticulum of membranes: their structure and possible interrelationship in *Chara corallina*. J Cell Biol 98:1537-1545

Pesacreta TC, Lucas WJ (1985) Presence of a partially-coated reticulum in angiosperms. Protoplasma 125:173-184

Sluiman HJ (1989) The green algal class Ulvophyceae. An ultrastructural survey and classification. Crypt Bot 1:83-94

Sluiman HJ, Blommers PCJ (1989) Ultrastructure and taxonomic position of *Chlorosarcina stigmatica* Deason (Chlorophyceae, Chlorophyta). Arch Protistenkd (in press)

Sluiman HJ, Kouwets FAC, Blommers PCJ (1989) Classification and definition of cytokinetic patterns in green algae: sporulation versus (vegetative) cell division. Arch Protistenkd (in press)

Sluiman HJ, Lokhorst GM (1988) The ultrastructure of cellular division (autosporogenesis) in the coccoid green alga, *Trebouxia aggregata*, revealed by rapid freeze fixation and freeze substitution. Protoplasma 144:149-159

Cytoskeletal dynamics in the reproductive cycle of *Acetabularia* (Dasycladaceae, Chlorophyceae): Cyst morphogenesis and delayed cytokinesis

Diedrik Menzel and Christine Elsner-Menzel
Max-Planck-Institut für Zellbiologie,
Rosenhof,
D-6802 Ladenburg,
F.R.G.

The marine green alga Acetabularia is in many ways a very unusual organism. Despite its giant dimensions of several centimeters in length and up to 0.5 mm in width, it is single-celled, containing for most of its 2 1/2 months life cycle a single giant nucleus. The cell undergoes tremendous changes in shape, commonly referred to as cellular morphogenesis (Werz, 1968b), which entail the formation of several storeys of slender, multi-membered whorls along the cylindrical stalk and the formation of an apical crown of larger rays. In several species of the genus Acetabularia, these rays may be laterally connected to form an umbrella-like structure called the cap (Valet, 1969).

Whereas most other cell types rely on informations sequentially issued from the nucleus for their morphogenesis, Acetabularia cells can undergo complex morphogenesis in the absence of a nucleus. Hämmerling, who introduced Acetabularia as a model system for the study of nucleo-cytoplasmic relationships, developed the concept that morphogenetic substances are formed by the nucleus and released into the cytoplasm very early in development (Hämmerling, 1963). More recently, it has become clear that these substances are equivalent to long-lived messenger RNAs (for summary, see Kloppstech, 1982). These and other unique features have made Acetabularia an attractive experimental system for studies on the control of morphogenesis (see Berger et al., 1987 and Bonotto, 1988 for most recent reviews).

Investigations on the morphogenesis of cysts at the end of the alga's life cycle have brought further interesting features to our attention. Among them is a series of cytoplasmic shape changes in the absence of cell wall formation resulting in the emergence of cyst protoplasts. Pioneering work by Werz and collaborators (Werz, 1965, 1968a, 1968b, 1969a, 1969b, 1970, 1974, Zimmer & Werz, 1981) and by Woodcock

(Woodcock 1971; Woodcock & Miller, 1973) provided preliminary evidence that cytoskeletal elements must be involved in these shape changes, long before the concept of a cytoskeleton in plant cells was generally accepted. Recently, immunocytochemical approaches have made it possible to directly visualize elements of the cytoskeleton and to explore their intracellular location throughout the life cycle in Acetabularia. The present account summarizes the work on cyst morphogenesis and then focuses on the dynamic behavior of the cytoskeleton in the transitions from one morphogenetic stage to the next.

Stages of cyst morphogenesis

Towards the end of the life cycle, the primary nucleus of Acetabularia shrinks considerably by extruding nucleolar material and undergoes meiosis by which the chromosome complement is reduced to 20 bivalents (Depesh N. De & Siegrid Berger, personal communication, Shihira-Ishikawa & Kuroiwa, 1984). Meiosis is immediately followed by many rounds of mitotic divisions in the absence of cytokinesis. Thus, the cell enters the coenocytic stage during which the thousands of secondary nuclei migrate up through the stalk into the apical cap ray chambers.

Cytokinesis finally occurs when the nuclei have settled in the cap rays, after a delay of more than a week. Prior to the actual cleavage process, a step by step reorganization in the cytoplasm takes place, creating new geometrical order independently of the symmetry axes originally established during the earlier mitoses. As a result of these rearrangements, mononucleate compartments are formed containing all the vital cell organelles needed to build the new cellular entities known as cysts.

While the secondary nuclei migrate up through the stalk, they are in the order of 5 μm in size and therefore not visible under the low power dissecting microscope. However, as soon as they arrive in the cap ray cytoplasm, they become visible as "white spots" (Fig. 1a-f). The nuclear volume increases by a factor of 10 and the nuclei become flatter. These changes are accompanied by the formation of a chloroplast-free, clear zone around the perimeter of the nuclei, resulting in a diameter for the "white spot" of approximately 25 μm.

As more and more nuclei appear in the cap ray cytoplasm over a period of about 24h, they gradually take on an orientation in linear rows, each nucleus carrying a small

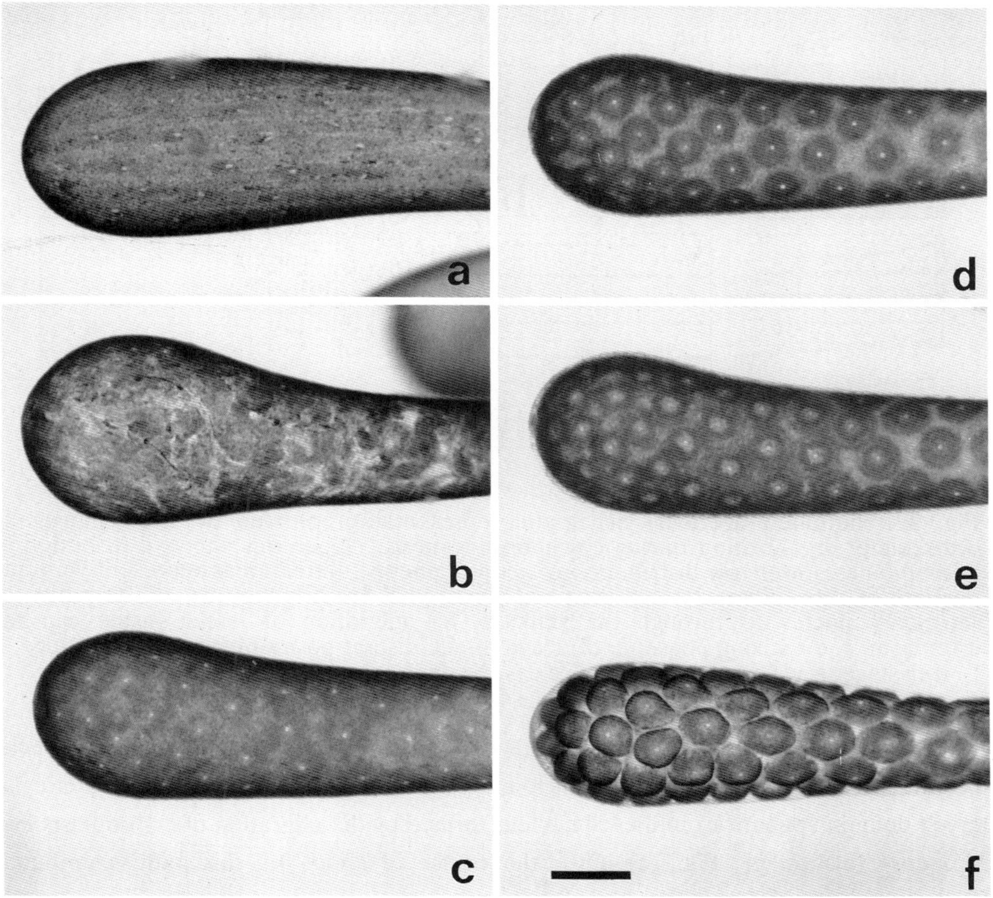

Fig. 1: Stages of cyst morphogenesis in cap ray chambers of Acetabularia cliftoni viewed under a low power dissecting microscope, scale 500μm; a) positioning of secondary nuclei in rows; b) mixing stage; c) equidistant positioning stage; d) disk stage; e) ring stage; f) dome stage; note apico-basal gradient in the morphogenetic sequence.

cluster of chloroplasts on either side in a lengthwise direction (Fig. 1a). Such rows may remain visible for about 12h before the cytoplasm takes on a rougher appearance due to the formation of many small clusters of chloroplasts dispersed through the cytoplasm (Fig. 1b). This change in texture marks the beginning of the "mixing phase" which is characterized by a peculiar reshuffling of nuclei particularly well perceivable by time lapse cinematography (Koop 1978).

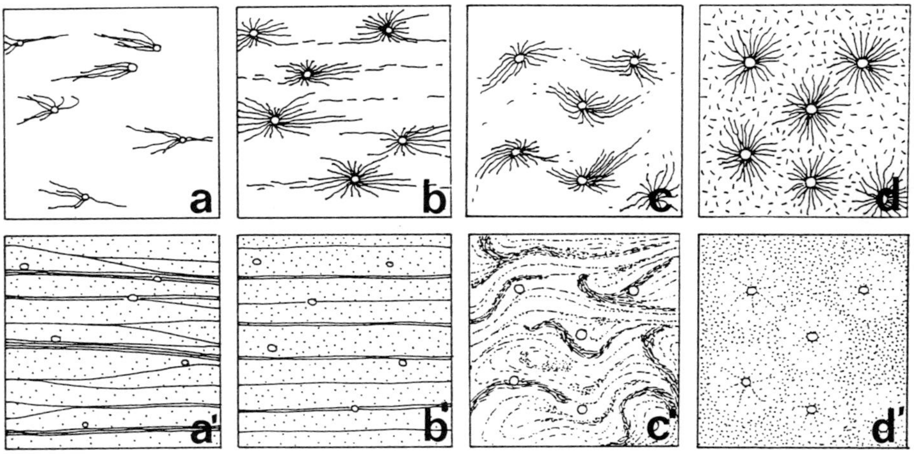

Fig. 2: (continued overleaf) Schematic representation of step by step changes in cytoskeletal organization during cyst morphogenesis. Upper row shows distribution of microtubules, lower row distribution of actin filaments; dotted areas represent very fine network of actin filaments; a-g' are front views, h, h' are side views. a, a') migration of secondary nuclei up through the stalk; b, b') migration of nuclei inside cap ray cytoplasm; c, c') mixing phase; d, d') stage of equidistant positioning; e, e') disk stage; f, f') ring stage; g, g') dome stage; h, h') cleavage stage.

Mixing may proceed for 8h before all movements subside and the secondary nuclei become positioned almost equidistant to each other (Fig. 1c). Concomitantly, a circular green domain appears around each nucleus formed by the migration of chloroplasts into this area (disk stage, Fig. 1d). Over the course of about 4h, this pattern gradually changes due to a transient outward transport of chloroplasts from the center of the domain to the periphery resulting in a green ring around each nucleus (ring stage, Fig. 1e). Subsequently, the outer edge of the ring becomes sharply defined and the domain starts to bulge outward resembling a miniature half-dome (dome stage, Fig. 1f). It is during this phase that the actual cleavage process takes place.

The remainder of the cyst-forming process is difficult to visualize from the original micrographs due to the large depth of field (this stage is schematically depicted in Fig. 2h, h'). Cleavage proceeds by a narrowing of the ring at the base of each domain which appears to sink inwards, accompanied by an increase of surface area over the half-dome so that a sphere is formed. The resulting spherical protoplasts finally detach from the residual mass of contracted material and differentiate into cysts. It is worthwhile noting

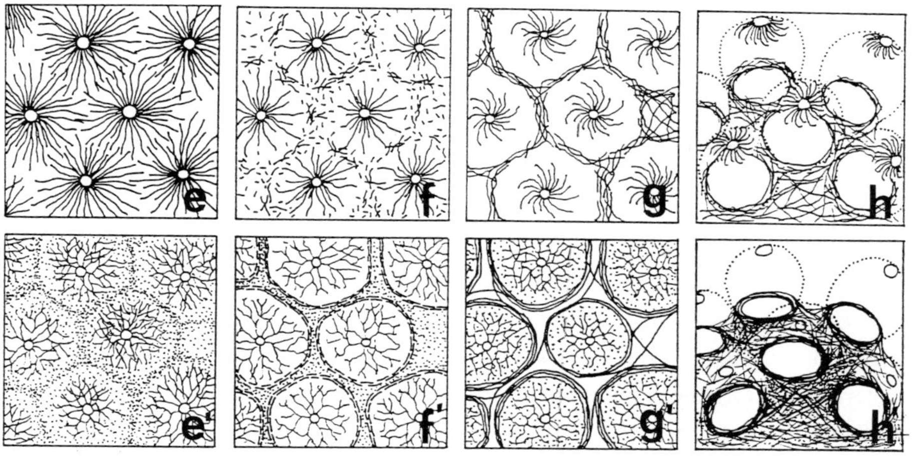

that the shape of the cyst protoplast varies among the species of Acetabularia and may be ovoid to oblong in some cases. This is a clear indication that control of shape is a function of the protoplasts and not a consequence of cell wall formation (Werz, 1968a).

Cytoskeletal changes accompanying cyst morphogenesis

The cytoskeleton in Acetabularia, as visualized by an indirect immunofluorescence procedure (Menzel, 1986), undergoes tremendous changes during the life cycle. These changes are schematically summarized in figure 2 and will be briefly described in the following (for a full description, see Menzel & Elsner-Menzel 1989a):

During the vegetative phase, prior to meiosis, an extensive system of parallel bundles of actin cables is present in the cytoplasm, running throughout the length of the stalk and cap rays (Menzel & Elsner-Menzel 1989), whereas microtubules are absent. Recent studies employing the western blot technique showed that total protein extracts from cells in stages prior to meiosis did not contain tubulin polypeptides, whereas tubulin was present in all subsequent stages up to gametogenesis (Elsner-Menzel & Menzel, 1989), indicating that tubulin expression is developmentally regulated.

Microtubules become visible for the first time in the meiotic spindle (Koop et al., 1979) and subsequently appear as a radial array of some 15 to 20 microtubules of variable length around each secondary nucleus as these organelles migrate through the stalk cytoplasm into the cap rays (Fig. 2a). Within the cap ray cytoplasm, a sudden increase in microtubule polymerization around the nuclear periphery occurs, resulting in an extension of the perinuclear microtubule systems over an area of 100 to 200 μm in diameter (Fig. 2b). During this phase the secondary nuclei arrange in linear rows as seen in figure 1a.

The next dramatic change takes place predominantly in the actin cytoskeleton. The regular arrays of parallel bundles gradually disintegrate (Fig. 2c') and give way to a network of finely interwoven actin fibrils (Fig. 2d'). During this process, the orderly movement of organelles, which has been prevalent during the entire vegetative period, also deteriorates. Organelles and the entire cytoplasm go through a phase of erratic back and forth movements and contractile events before all movements subside. During this phase, the nuclei become mixed as shown above (Fig. 1b). At the end of this period, changes are also evident in the microtubule cytoskeleton by the appearance of many small microtubule fragments in the cytoplasm, indicative of a change in the regulation of microtubule initiation which was restricted to the perinuclear area up to this developmental stage (Fig. 2d).

These microtubule fragments, however, do not elongate further to give rise to additional microtubules outside the perinuclear systems but become incorporated into the perinuclear systems. As the cells pass from the equidistant positioning stage to the disk stage, these microtubule systems reach their greatest extension whereby adjacent systems partly overlap (Fig. 2e). During this stage, the actin cytoskeleton remains undifferentiated in the inter-domain area, whereas very fine radial filaments appear around the nuclei (Fig. 2e').

Gradually, the actin network in the inter-domain spaces reorganizes; filament bundles appear and become grouped into parallel ring segments (Fig. 2f'). Finally, an entire ring of actin filaments is formed around the perimeter of each domain. At the same time, microtubules start fragmenting at their distal ends and become oriented in parallel with the forming actin ring (Fig. 2f). This phase is equivalent with the ring stage shown in figure 1e.

The actin rings further rearrange into massive, very regular actin bands, separated by a honeycomb-patterned network of microtubules which is termed the peripheral microtubule system (Fig. 2 g, g'). With the beginning of the dome stage, the actin rings start decreasing their diameters, which may be interpreted as a contraction event

(Menzel 1988), and microtubules in the peripheral system become compressed and eventually undergo depolymerization. The perinuclear, radial microtubule arrays, on the other hand, decrease in diameter and the microtubules become uniformly bent counter-clockwise to form whirls. This could be the consequence of the contraction process leading to minute, rotationally directed rearrangements in the plane of the plasma membrane to which the perinuclear microtubule arrays appear to be attached (Fig. 2h & h').

This brief summary of events clearly demonstrates that the actin cytoskeleton, as well as the microtubule cytoskeleton, continuously rearrange during cyst morphogenesis. It may be hypothesized that these rearrangements are regulated by factors governing polymerization and depolymerization dynamics of the two filament systems and modes of interactions such as crosslinking, lateral association, and directional force production. It is further conceivable that in order for the cell to pass from one stage to the next new factors need to be added in a sequential manner to the preexisting cytoskeleton and that these new components change not only the structure of the cytoskeleton but also its dynamic behavior.

Arrest of cyst morphogenesis

By applying inhibitors of DNA-, RNA-, and protein synthesis, we have sought to determine whether newly synthesized proteins are involved in the transitions from one morphogenetic stage to the next and whether transcriptional activities are involved or replication of DNA is required. Seven phases were selected for the placement of pharmacological blocks (I-VII in Fig. 3), and three inhibitors were applied independently: Aphidicolin (inhibits DNA-polymerase, (Ikegami et al., 1978, Spadari et al., 1985), 2,6-DCBR (2,6-dichlorobenzimidazole riboside, inhibits RNA-Polymerase-II, Tamm et al., 1976, Zandomeni et al., 1986), and cycloheximide (inhibits the peptide transfer reaction on 80S ribosomes, Ennis & Lubin, 1964, Pestka, 1977).

All three inhibitors block the transition from the migratory stage of secondary nuclei to the initiation of cyst morphogenesis (Block I, Fig. 3). This result suggests that synthesis of new proteins is necessary for the transition and that the information for these new proteins needs to be freshly transcribed from the nuclear genome. Furthermore, the inhibition by aphidicolin indicates that if progression through the nuclear division cycle is blocked in the S-phase subsequent retrieval of genetic information via transcription/translation is also inhibited.

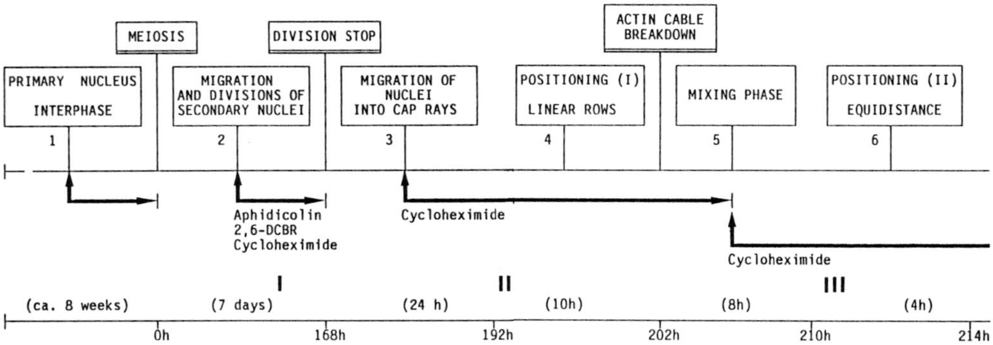

Fig. 3: (continued overleaf) Diagram depicting the linear sequence of morphogenetic steps along the time scale. Upper row boxes points out major events in the life cycle of Acetabularia, lower row boxes contain the stage names. Approximate start and end of each stage is given in hours. Meiosis is arbitrarily chosen as 0h. The time intervals in brackets above the time scale indicate the duration of each stage. Intervals of inhibitor treatments are indicated and numbered I-VII.

During normal development, secondary nuclei which have reached the cap ray cytoplasm cease to divide as they start to organize the perinuclear microtubule system around them. Neither aphidicolin nor 2,6-DCBR have an effect on cyst morphogenesis at this or any of the following steps in morphogenesis up to the stage when cyst protoplasts have developed and the single cyst nucleus starts to divide. This clearly indicates that all messages necessary for the morphogenetic sequence during cyst formation are present in the cytoplasm and further transcriptional activity is not required. However, the first cyst mitosis and all the following mitoses of the cyst nuclei are strongly affected by aphidicolin, but surprisingly not by 2,6-DCBR (Block VII, Fig. 3). One explanation for this behavior may be that, whereas aphidicolin blocks DNA reduplication and thus affects the nuclear division cycle in the cyst, all messages for the cyst-specific processes have already been transcribed before the first mitotic division starts and, hence, an inhibition of transcription by 2,6-DCBR does not stop nuclear division.

On the other hand, continuous or stagewise translation of mRNAs into proteins is required during the morphogenesis of cysts (Blocks II-VI, Fig. 3). Application of cycloheximide during the initiation stage, when secondary nuclei have arrived in the cap rays, does not block the transition into the mixing stage, inferring that limited actin cable breakdown, which is typical for the mixing stage, can take place in the absence of protein synthesis. However, complete breakdown of actin cables, reorganization of the actin cytoskeleton into an irregular, fine network and equidistant positioning of nuclei do not

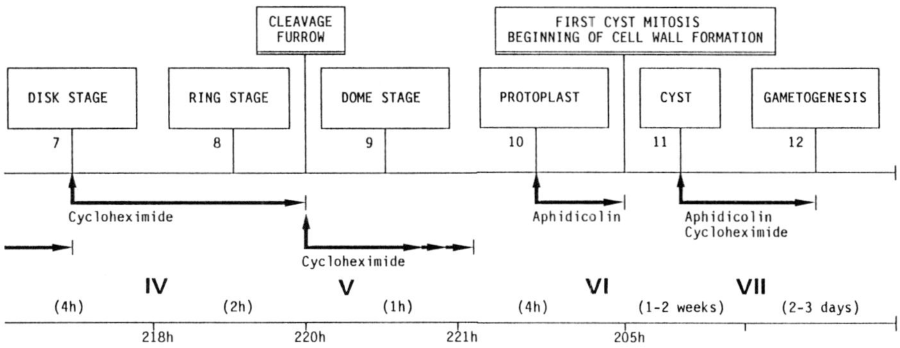

take place. Therefore, transition from the mixing stage into the next stage is blocked (Block II, Fig. 3). If application of cycloheximide is continued, the mixing stage also persists for several days before detrimental effects set in (Fig. 4a).

If the cells are allowed to proceed unaffected into the mixing stage and cycloheximide is then applied, positioning of nuclei eventually occurs although the distances between nuclei are much more irregular. The cells even enter into the disk stage but further development becomes aberrant and eventually ceases (Block III, Fig. 3, Fig. 4 b & c).

Cells which are incubated in cycloheximide at the beginning of the disk stage may proceed into the ring stage but the formation of the cleavage furrow fails so that these cells cannot pass into the dome stage (Block IV, Fig. 3). Finally, cells which have made it halfway through the dome stage displaying well developed cleavage furrows appear to regress into an undifferentiated state if cycloheximide is added (Block V, Fig. 3). This is most evident by a flattening of the cleavage furrows and a disappearance of domain borders (Fig. 4 e & f).

These findings support the idea that at several points along the morphogenetic sequence from the multinucleate cap ray cytoplasm to spherical uninucleate cyst protoplasts the appearance of new proteins is required in order to allow a transition from one stage to the next. It is not yet clear, however, how many such transition points exist. First, it needs to be established how long it takes for the drug to diffuse into the cell and, secondly, how quickly protein synthesis is suppressed. An apparent delay of inhibition could result from the existence of a pool of proteins that need to be turned over before inhibition of protein synthesis can take effect.

Effects of cycloheximide on the cytoskeleton

Immunofluorescent visualization of the cytoskeleton of cells blocked at various points during cyst morphogenesis reveals a strinkingly different effect of cycloheximide on the microtubule cytoskeleton on the one hand and the actin cytoskeleton on the other.

Up to the mixing stage, the actin cytoskeleton looks indistinguishable from its normal appearance as shown earlier (Menzel & Elsner-Menzel, 1989a). However, in the post-mixing stages the actin cytoskeleton falls into an unorganized network-like configuration about 1-2 h after the addition of cycloheximide. This unorganized state of the actin cytoskeleton resembles the configuration seen normally at the transition point between the mixing stage and the equidistant positioning stage. It must be concluded, therefore, that the presence of components for building up an organized array of actin rings is constantly required at the end of the mixing stage and during all subsequent stages.

For the microtubule system, the situation is different. Under normal conditions, microtubules are exclusively organized by the nuclear envelope (or a layer of cytoplasm closely around it) and remain associated with it (Menzel & Elsner-Menzel, 1989a). This contrasts markedly with the situation after cycloheximide application (Block II, Fig. 3) when microtubules appear at a very high density throughout the entire cytoplasm. Interestingly, the principle organization of the perinuclear microtubule arrays is not overly distorted when compared to that of untreated plants (Fig. 5a, compare with Fig. 5 in Menzel & Elsner-Menzel, 1989a).

It is noteworthy that the additional microtubules are not randomly distributed but organized in parallel with the overall lengthwise striation of the cytoplasm (Fig. 5b). This could either be due to a strong self-organizing capacity of the microtubules or to specific interactions with yet undetermined components in the cytoplasm which themselves are distributed regularly. At the beginning of the cycloheximide block-II, parallel actin cables are still present and distributed exactly in the same order, although they do not persist in the presence of cycloheximide. Therefore, initially, they are good candidates for providing a spatial template for the microtubules as they appear under the influence of cycloheximide.

During the course of cycloheximide block-III, which is characterized by the aberrant disks as seen in figure 4b, the perinuclear microtubule systems attain an asymmetric configuration with a large semicircular mass of entangled microtubules on

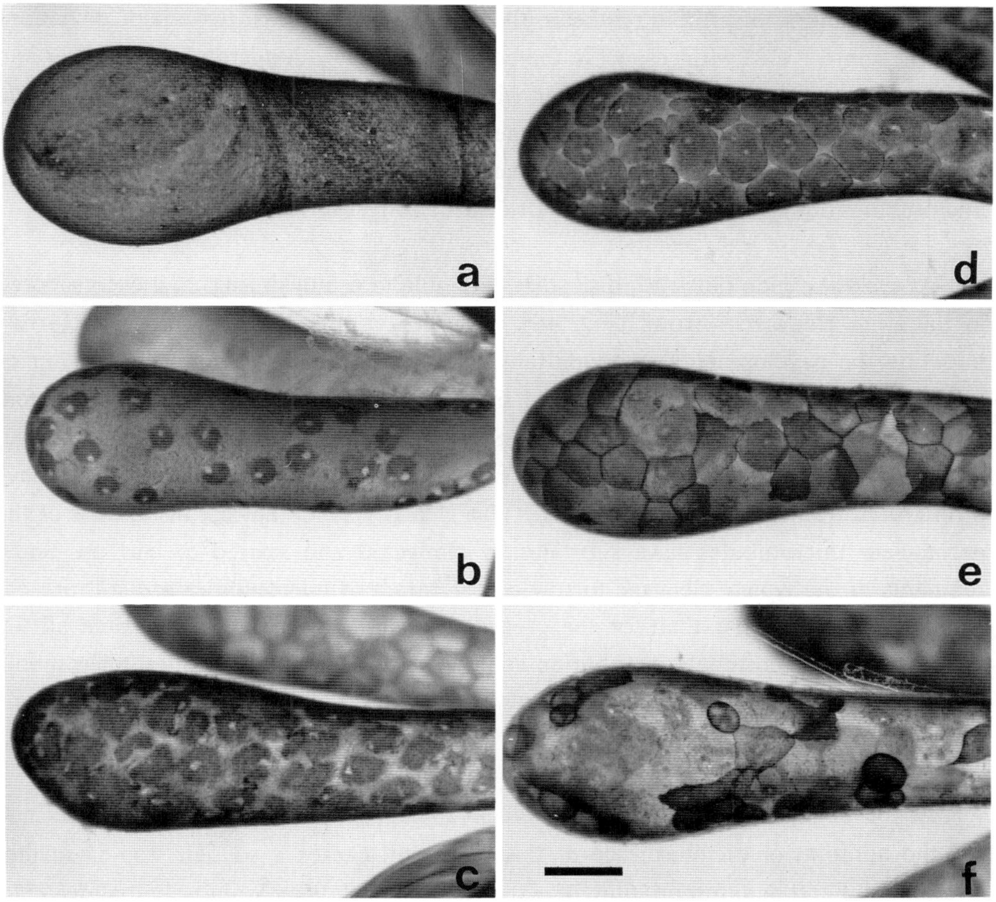

Fig. 4: Aberrant morphogenesis during cycloheximide blocks (scale 500μm).
a) Appearance of cap ray cytoplasm arrested in the mixing stage (Block II), 24 h treatment; b) cap ray arrested in the disk stage (Block III), 4h treatment; c) same stage as (b) but 8h treatment; d) cap ray treated with cycloheximide at the ring stage (Block IV), cleavage furrow is initiated but domains fail to protrude, 2h treatment; e) same cap ray as in (d) after another hour of treatment, domains begin to dedifferentiate; f) cap ray during block V, cleavage furrows have initially been formed but deteriorated quickly, domain structure is lost over large areas of cytoplasm, 1h treatment.

one side of the nucleus, whereas the opposite side looks less distorted (Fig. 5c). The area containing the entangled mass of microtubules is equivalent to the irregular disk area seen in the dissecting microscope (Fig. 4b). Many additional microtubules are scattered throughout the remainder of the cytoplasm. Obviously, the parallel order seen in the previous stage is lost in this population of microtubules.

The inhibition of protein synthesis, therefore, causes at least two major deviations from the control. One with respect to the equilibrium between microtubule polymer and tubulin monomer pools and the other with respect to the mode of interaction and alignment between microtubules and other components and/or among microtubules themselves. These deviations in turn cause not only an uncurbed proliferation of microtubules but also a difference in their distribution patterns.

At cycloheximide block-VI, disk stages are able to establish rings and even succeed in initiating cleavage furrows (Fig. 4d). However, at the onset of cleavage furrow initiation the developmental program fails. The cytoskeletal changes occurring during this failure are quite similar to those of the previous blocks. The actin cytoskeleton, which by that stage in normal algae shows the characteristic actin rings, has now de-differentiated into a random network (not shown) and microtubules again have proliferated tremendously (Fig. 5d). It is also clearly visible, that the distal ends of the perinuclear microtubules have substantially elongated and started to curl up. Some remnants of the typical cleavage furrow configuration of microtubules are still present (Fig. 5d, center). Other than for the preceding cycloheximide block, this time all the changes in the microtubule distribution could be ascribed to uncontrolled microtubule proliferation. The curling of the distal ends of perinuclear microtubules may simply be a consequence of spatial restraints.

At the beginning of cycloheximide block-V, cleavage furrows have already formed but the addition of cycloheximide at this point still causes a sudden stop in further development and dedifferentiation of cyst domain morphology as described above (Fig. 4 e & f). The changes occurring in the configuration of microtubules again are very similar to those described for the previous cycloheximide block (Fig. 5d), and the actin cytoskeleton again completely de-differentiates (not shown).

Fig. 5: Distribution of microtubules in cap rays treated with cycloheximide (scale 100μm). a) Extreme proliferation of microtubules in addition to the perinuclear microtubule systems during block II; b) close up of parallel arrangement of microtubules in the cytoplasmic network, such a distribution of microtubules is absent from untreated controls; c) distribution of microtubules in the disk area during block III, note that only half of the perinuclear microtubule system takes part in the formation of the aberrant disk, the other half radiates off freely; d) microtubule distribution during block IV and V, perinuclear microtubules are extremely long and curled up at their distal ends.

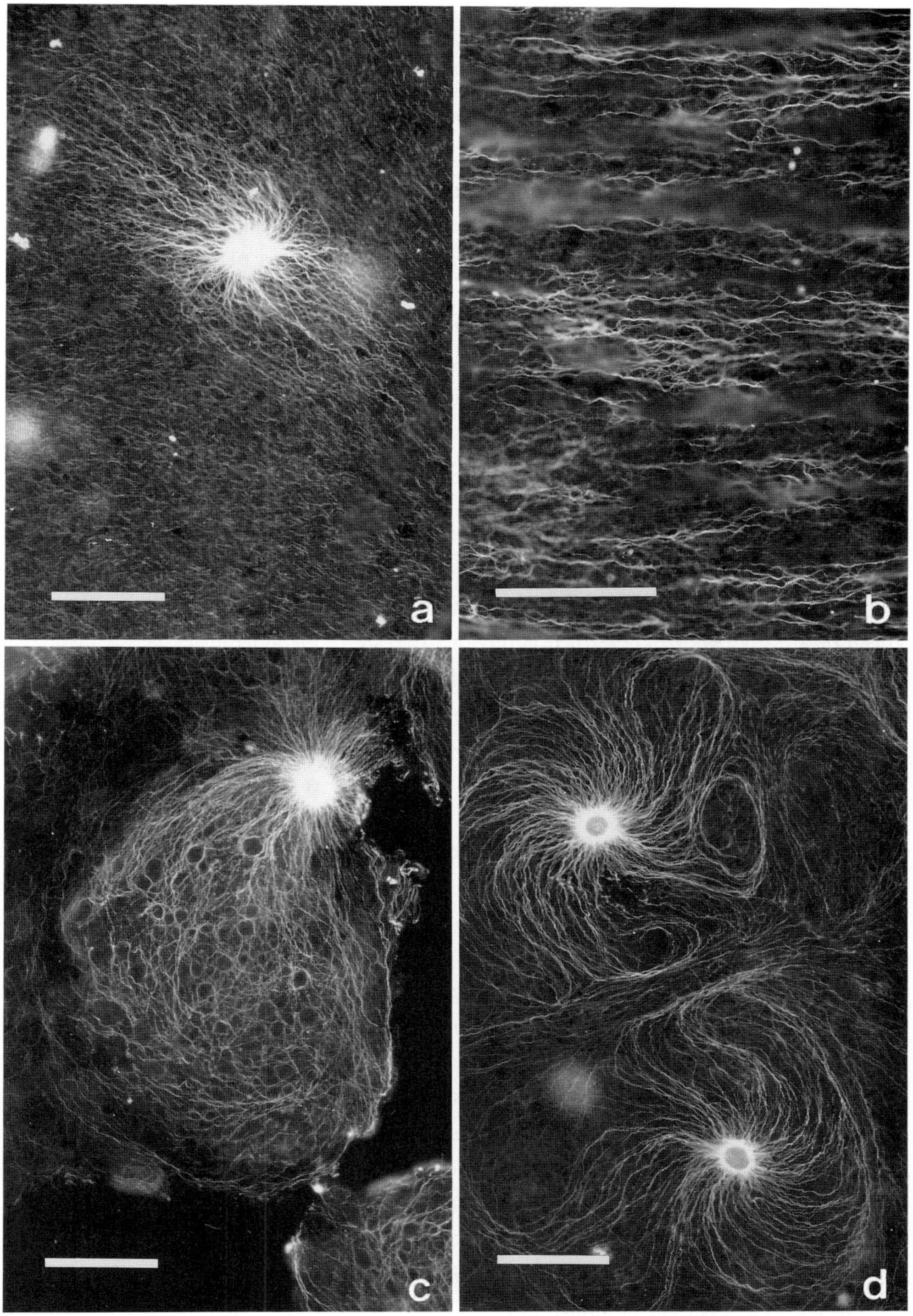
a
b
c
d

Conclusions

During vegetative growth of coenocytic green algae, the uncoupling of mitosis from cytokinesis leads to the coexistence of many nuclei in the same cell. However, at some point later in the life cycle cytokinesis must occur if uninucleate reproductive cells are to be formed. Usually, symmetry axes and cleavage planes are predefined by the symmetry of the mitotic apparatus, but how do these organisms establish division planes de novo? The uninucleate giant green alga Acetabularia is a particularly good model system to study this question. It turns out that in this alga information of the spatial order, necessary for cytokinesis, is lost after meiosis of the giant primary nucleus. However, after many cycles of mitotic divisions spatial order becomes reestablished prior to cytokinesis through the dynamics of the cytoskeleton.

We have previously suggested (Menzel & Elsner-Menzel, 1989a) that the radial, perinuclear microtubule systems play a leading role in the process of stepwise establishment of symmetry and that all subsequent steps, i.e., the definition of domain borders by circular bands of actin bundles are a consequence of the new symmetry. We have also demonstrated that both microtubules and actin filaments are operating coordinately in these processes (Menzel, 1988). It is obvious that cyst morphogenesis passes through a step by step series of changes and that each step sets the stage for the next one, eventually terminating in cytokinesis. Here, we have approached the question of whether stage-specific expression of proteins is involved in accomplishing the transition from one step to the next.

It is evident that the failure of completing cyst morphogenesis in the presence of cycloheximide is not a consequence of a lack of tubulin or actin. These two major cytoskeletal proteins should be present in vast pools from the beginning of cyst morphogenesis throughout the entire sequence of morphogenetic steps. Therefore, inhibition by cycloheximide can only be caused by the lack of proteins which regulate cytoskeletal dynamics. Regulation could be effective at least on two different levels, on the level of cytoskeleton-modifying proteins (e.g. MAPs, actin-associated proteins) or on the level of signalling proteins (e.g. calmodulin-regulated proteins, kinases, etc). At present, we have no way of defining which of the two levels is actually affected, but the immunocytochemical data of cytoskeletal changes during cycloheximide blocks clearly help, to narrow down which aspect of the cytoskeletal dynamics is changed by the absence of stage-specific proteins.

In all stages tested, cycloheximide causes the rearrangement of the actin filament cytoskeleton from ordered arrays into random networks. It appears to be significant that at the transition from the mixing stage to the stage of equidistant positioning of nuclei the actin cytoskeleton takes on a random network configuration naturally. We infer that this is a configuration of least complexity which serves as starting material for all subsequent higher order actin filament configurations and that, in order to rearrange this, the cell needs actin-associated proteins to specifically regulate filament length and interactions between actin filaments. Since the actin cytoskeleton falls back into the random network configuration during all cycloheximide blocks, even from the highly elaborate arrangements during the dome stage, it must be assumed that these proteins are turned over quickly and need to be constantly expressed to maintain an operational level.

The changes occurring in the microtubule cytoskeleton during cycloheximide blocks are quite different from those in the actin cytoskeleton. However, all cycloheximide blocks have one characteristic in common: the uncurbed proliferation of microtubules. Hence, it may be concluded that tubulin polymerization is tightly controlled by proteins which are probably also turned over quickly and are constantly replenished by protein synthesis, for the same reasons as discussed above in the case of actin-associated proteins. In addition to changes in the polymerization dynamics, changes also occur in microtubule distribution. Some of these changes can be attributed to space constraints and may therefore be of secondary significance. Others, however, should be due to the lack of proteins which are involved in regulating microtubule-microtubule binding and interaction of microtubules with other cell components such as chloroplasts. This is particularly evident during cycloheximide block-III when microtubules are becoming lumped together in irregularly shaped disks containing the majority of the chloroplasts.

Outlook

In contrast to the situation in animal cells, where much has been learned in the recent past on the polymerization dynamics of actin and microtubules and the role of accessory proteins in cytoskeletal interactions, little is known about these processes in plant cells. One reason for this lack of understanding is that only a limited number of plant cell systems display these phenomena clearly enough to be experimentally approachable. Our observations show that Acetabularia is certainly one of the favorable model systems offering an unusually large size, a highly dynamic cytoskeleton, and stage-specific expression of cytoskeleton-modifying/regulating proteins. It should therefore be expected that future progress in the understanding of cytoskeletal dynamics in plant cells will be stimulated by the exploitation of this combination of unique features.

Acknowledgements

We would like to express our sincere gratitude to Prof. Peter Traub for continued support and encouragement. We also thank Dr. Julia Willingale for careful reading of the manuscript and Mrs. Heidi Klempp for secretarial assistance. Financial support by the Deutsche Forschungsgemeinschaft for part of this work is acknowledged.

References

Berger, S, de Groot, EJ, Neuhaus, G, Schweiger, M (1987) Acetabularia: A giant single cell organism with valuable advantages for cell biology. Eur J Cell Biol 44:349-370.

Bonotto, S (1988) Recent progress in research on Acetabularia and related Dasycladales. Progress in Phycological Research 6:59-235.

Elsner-Menzel, C, Menzel, D (1989) Cytoskeletal actin and tubulin in the siphonous green algae Acetabularia, Bryopsis, and Caulerpa. Eur J Cell Biol 48 (Suppl 26):75.

Ennis, HL, Lubin, M (1964) Cycloheximide: aspects of inhibition of protein synthesis in mammalian cells. Science 146:1474-1476.

Hämmerling, J (1963) Nucleo-cytoplasmic interaction in Acetabularia and other cells. Ann Rev Plant Physiol 14:65-92.

Ikegami, S, Tagichi, T, Ohashi, M, Nagano, H, Mano, Y (1978) Aphidicolin prevents mitotic cell division by interfering with the activity of DNA polymerase alpha. Nature 275:458-460.

Kloppstech, K (1982) Acetabularia. In: Smith H, Grierson D (eds.), The Molecular Biology of Plant Development. Blackwell Scientific Publications. Oxford, London. p 136-158.

Koop, HU (1978) Entwicklung von Acetabularia (Dasycladales). Inst fd Wiss Film, Göttingen, Film C1298.

Koop, H-U, Schmid, R, Heunert, H-H, Spring, H (1979) Spindle formation and division of the giant primary nucleus of Acetabularia (Chlorophyta, Dasycladales). Differentiation 14:135-146.

Menzel, D (1986) Visualization of cytoskeletal changes through the life cycle in Acetabularia. Protoplasma 134:30-42.

Menzel, D (1988) Perturbation of cytoskeletal assemblies in cyst domain morphogenesis in the green alga Acetabularia. Eur J Cell Biol 46: 217- 226.

Menzel, D, Elsner-Menzel, C (1989a) Maintenance and dynamic of cytoplasmic organization controlled by cytoskeletal assemblies in Acetabularia. In: Stein, J, Coleman, AW, Goff, LJ, (eds.). Algae as Experimental Systems in Cell Biology. Alan R Liss, p. 71-91.

Menzel, D, Elsner-Menzel, C (1989b) Induction of actin-based cytoplasmic contraction in the siphonous green alga Acetabularia (Chlorophyceae) by locally restricted calcium influx. Botanica Acta 102:164-171.

Pestka, S (1977) Inhibitors of protein synthesis. In: Weissbach, H., Pestka, S. (eds.), Molecular mechanisms of protein biogenesis. Academic Press, NY, SF, London. p 467-553.

Shihira-Ishikawa, I, Kuroiwa, T (1984) Morphological transition of the nucleus during the whole life cycle of Acetabularia calyculus QUOY et GAIMARD. Jap J Phycol 32:147-157.

Spadari, S, Focher, F, Sala, F, Ciarrocchi, G, Koch, G, Falaschi, A, Pedrali-Noy, G (1985) Control of cell division by aphidicolin without adverse effects upon resting cells. Drug Res 35:1108-1116

Tamm, I, Hand, R, Calguiri, LA (1976) Action of dichlorobenzimidazole riboside on RNA synthesis in L-929 and HeLa cells. J Cell Biol 69: 229- 240.

Valet, G (1969) Contribution a l'étude des Dasycladales 2. Cytologie et reproduction. 3. Révision systématique. Nova Hedwigia 17:551-644.

Werz, G (1965) Determination and realization of morphogenesis in Acetabularia. Brookhaven Symp Biol 18:185-203.

Werz, G (1968a) Plasmatische Formbildung als Voraussetzung für die Zellwandbildung bei der Morphogenese von Acetabularia. Protoplasma 65:81-96.

Werz, G (1968b) Differenzierung und Zellwandbildung in isoliertem Cytoplasma aus Acetabularia. Protoplasma 65:349-357.

Werz, G (1969a) Wirkung von Colchicin auf die Morphogenese von Acetabularia. Protoplasma 67:67-78.

Werz, G (1969b) Morphogenetic processes in Acetabularia. In: Bücher, T., Sies, H. (eds.). Inhibitors - Tools in Cell Research. Springer Verlag, Berlin, Heidelberg, New York. p 167-186.

Werz, G (1970) Mechanism of cell wall formation in Acetabularia. In: Brachet, J., Bonotto, S. (eds.). Biology of Acetabularia. Academic Press New York, London. p 125-144.

Werz, G (1974) Fine structural aspects of morphogenesis in Acetabularia. Int Rev Cytol 38:319-367.

Woodcock, CLF (1971) The anchoring of nuclei by cytoplasmic microtubules in Acetabularia. J Cell Sci 8:611-621.

Woodcock, CLF, Miller, GF (1973) Ultrastructural features of the life cycle of Acetabularia mediterranea. II. Events associated with the division of the primary nucleus and the formation of cysts. Protoplasma 77: 331-341.

Zandomeni, R, Zandomeni, MC, Shugar, D, Weinmann, R (1986) Casein kinase type II is involved in the inhibition by 5,6-dichloro-1-beta-ribofuranosylbenzimidazole of specific RNA polymerrase II transcription. J Biol Chem 261:3414-3419.

Zimmer, B, Werz, G (1981) Cytoskeletal elements and their involvement in Polyphysa (Acetabularia) protoplast differentiation. Cytoskeleton modifiers and concanavalin A-mediated effects. Exptl Cell Res 131:105-113

Notes on heterokaryotic life cycle phases in some Dasycladales and Bryopsidales (Chlorophyta)

R. Schnetter, B. Brück, K. Gerke, and G. Seibold
Botanisches Institut I and Strahlenzentrum
Justus-Liebig-Universität
D-6300 Giessen
FRG

The siphonous marine Chlorophyta are comprised of the orders Dasycladales and Bryopsidales. Convincing information on the behaviour of the nuclei during the complete life cycle in the Dasycladales exists so far only for *Acetabularia* Lamouroux (for literature see Bonotto, 1988). The relatively long diplophase, starting with zygote formation, is followed by a heterokaryon beginning after the meiosis of the primary nucleus. Secondary nuclei undergo unsynchronous mitotic divisions before cyst formation. The latter starts with a single nucleus dividing mitotically within the young cyst. Therefore, the cysts are homokaryotic and release male or female gametes. Nuclear behaviour in *Batophora* J. Agardh (Liddle et al., 1976) is probably similar to that in *Acetabularia*, but information on the ploidy levels of the nuclei is still lacking for *Batophora.*

Reviewing the literature on life cycles in the Bryopsidales, Tanner (1981) states that, with the exception of the Bryopsidaceae, all members of this order are assumed to have a diplontic life cycle with a gametic meiosis. Unfortunalety, karyological studies are nearly completely lacking in most Bryopsidales. For the Bryopsidaceae, a diplontic life history was presumed until Kornmann (1938) found that the spherical *Halicystis ovalis* (Lyngb.) Areschoug is the gametophytic phase of the filamentous *Derbesia marina* (Lyngb.) Solier. The latter is formed after gametic mating and produces stephanokontan spores. Germlings of zoospores develop into gametophytes. A corresponding relationship between *Halicystis parvula* Schmitz and *Derbesia tenuissima* (De Not.) Crouan was observed by Feldmann (1950). Since then, it has been generally accepted that the gametophytic *Halicystis* thalli, which produce macrogametes or microgametes, are haploid and that the filamentous and spore-producing *Derbesia* plant should be diploid. This concept was strongly supported by the observation that meiosis takes place in the sporangia of *Derbesia marina* and *D. tenuissima* (Neumann, 1967, 1969b). Due to their small size, chromosomes could only be observed and counted in the sporangia during meiosis (Neumann, 1969b).

Ploidy level studies in a Colombian strain of *Derbesia tenuissima* by microspectrophotometry (after Feulgen staining or staining with the Hoechst dye 33258) showed, in contrast to this concept, that only haploid nuclei are present in the filaments of sporophytic thalli (Schnetter et al., 1981). Initially, this observation was considered as an exception, but identical results were obtained in further strains of *Derbesia tenuissima* as well as in *D. marina* and in *D. cf. novaezelandiae* Chapman (Schnetter et al., 1985; Eckhardt et al., 1986). Studies on nuclear behaviour in the fusion product of micro- and macrogametes have shown that, instead of karyogamy, both gamete nuclei divide mitotically (Eckhardt & Schnetter, 1984). High nuclear DNA contents in *Derbesia tenuissima* from Rovinj (Yugoslavia), corresponding to diploid levels, could only be observed microspectrophotometrically after Feulgen

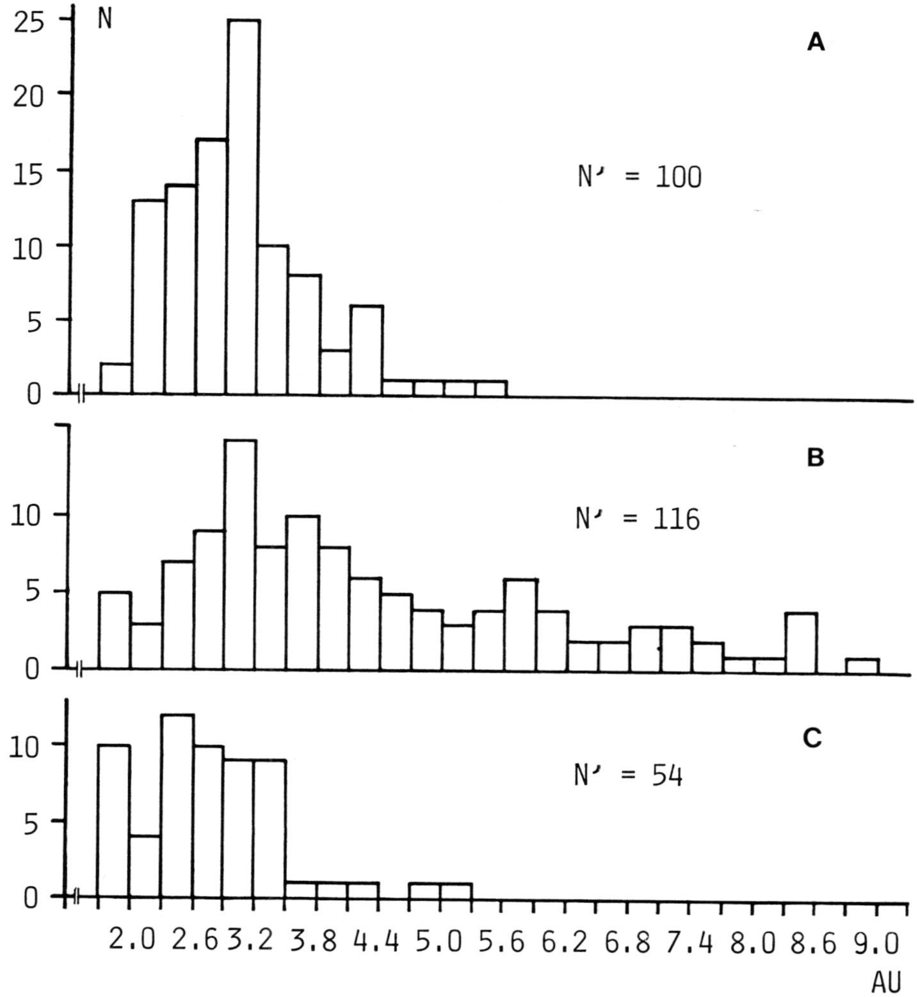

Fig. 1 : Relative nuclear DNA content of filaments of sporophytes (A), young sporangia still in plasmatic connection with the supporting filament (B), and older sporangia (C) of *Derbesia marina*, measured in arbitrary units (AU) by Feulgen microspectrophotometry (see Al-Kubaisy et al., 1981; Schnetter et al., 1985). N - number of nuclei; N - total number of measured nuclei.

staining in young sporangia still in protoplasmic contact with the main filament (Eckhardt et al., 1986). Sporophytes obtained parthenogenetically from macrogametes do not show higher nuclear DNA contents in their sporangia. It is therefore assumed that karyogamy will only take place when both female and male nuclei are present. Moreover, in a *Derbesia marina* isolated from Helgoland, nuclei with higher DNA content, indicating diploidy, only occur in young sporangia (Fig.1).

At this point the question arises, does a heterokaryon occur in other genera of the Bryopsidaceae? One isolate, collected at the Spanish Atlantic coast, was considered to belong to *Derbesia marina* due to the shape of the zoosporangia, the "double septum" generally present within their bases, and the absence of pyrenoids in the chloroplasts. Germlings raised from zoospores develop into irregularly growing filaments (Eckhardt et al., 1986, fig. 3, as *Derbesia marina*). In contrast to the irregularly growing germlings of *Halicystis ovalis* obtained by Rietema (1973), variations of culture conditions did not influence the shape of the germlings of the Spanish isolate. After culturing it for about 7 years, the rugose filaments developed discoidal structures typically for the genus *Pedobesia* MacRail & Womersley. The collection site of our isolate and the thallus characters indicate that our plant is *Pedobesia solieri* Abelard & Knoepffler. The plug structure within the base of the sporangium can no longer be considered a distinguishing character between *Derbesia* and *Pedobesia*, as stressed by several authors (see Abelard, 1982, 1986; Abelard & Knoepffler, 1986). The separating structure within the base of sporangia in *Derbesia tenuissima* consists of a plug and a thin cell wall oriented towards the filament (Menzel, 1980). We found that in cultured isolates of *Derbesia tenuissima*, *D. marina*, as well as in *Pedobesia solieri*, this plug-wall complex is identical, if studied with the light microscope, but is highly variable in its expression (see also Menzel, 1980, fig. 8; Abelard & Knoepffler, 1986, fig. 18). In some cases, a thick cell wall has been observed in *Pedobesia solieri*, formed by the supporting filament below the plug-wall complex of old sporangia (Fig. 2).

In *Pedobesia solieri*, comparisons of nuclear DNA contents show that, as in *Derbesia*, nuclei with increased DNA only exist in very young sporangia still in plasmatic connection with the supporting filament. Thus it may be assumed that, with the exception of the nuclei with higher DNA contents in young sporangia, all nuclei of *Pedobesia* are haploid. Karyogamy occurs in young sporangia immediately followed by meiosis (Eckhardt et al., 1986, fig. 8, material from Celorio).

Gametangia and gametes have never been observed in *Pedobesia* (MacRail & Womersley, 1974; Kobara & Chihara, 1984; De Paula & West, 1986). Nevertheless, zoospore germlings of *Pedobesia* have some characters typical for *Derbesia* gametophytes: the germlings are homokaryotic, because the initial spore stages in the sporangia formed after karyogamy and meiosis are uninucleate. Staining characteristics of the cell wall correspond to those of *Halicystis* (Rietma, 1975), indicating xylan and cellulose.

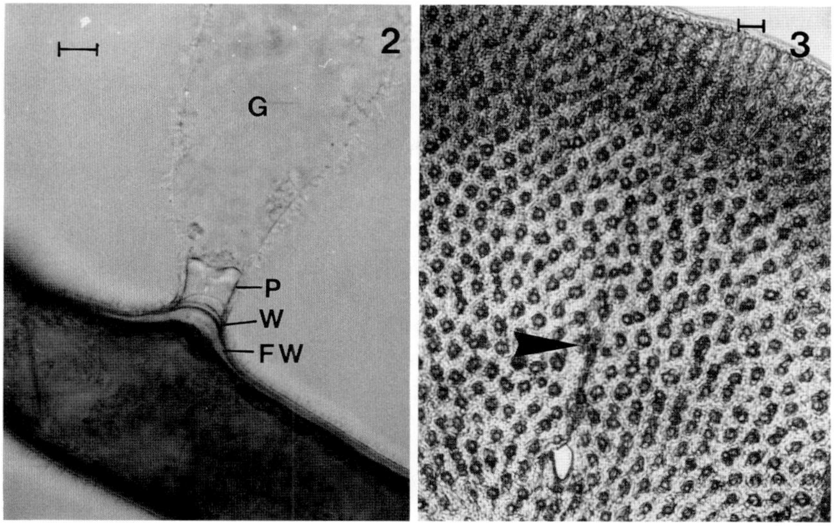

Figs. 2,3 : *Pedobesia solieri.* Fig. 2 : Empty sporangium showing ghost (G), plug (P) - wall (W) complex, and thick wall formed by the supporting filament (FW). Fig. 3: Disk showing anastomosis of two lobes (arrowhead) and posterior adjustment of the symmetric thallus structures. Scales : 10 um.

In our cultures, separately cultivated germlings of zoospores never developed the smooth, sporangia-bearing filaments that were collected. We suppose that the smooth, sporangia-bearing filaments of *Pedobesia* are also heterokaryotic, because karyogamy in *Derbesia tenuissima* only takes place in the sporangia of heterokaryotic sporophytes. Perhaps, the absence of gametes in *Pedobesia* indicates an up to date unknown form of somatogamy. We observed that disks of one plant are able to fuse. The discoidal structure is then completely adjusted (Fig. 3).

The sporophytes of *Bryopsidella* Feldmann species have sporangia of the *Derbesia* type. Rietma (1972) supposed that the sporophyte of *Bryopsidella neglecta* (Berthold) Feldmann in Rietema (Syn.: *Derbesia neglecta*, Berthold/*Bryopsis halymeniae* Berthold) has diploid nuclei, but he only counted the chromosomes of the gametophyte and observed meiosis in the sporangium before plug formation. These observations do not exclude the possibility of the existence of a heterokaryon in the sporophytic filaments. Measurements of nuclear DNA contents in *Bryopsidella ostreobiformis* Calderon-Saenz & Schnetter, a Mediterranean species with nearly isomorphic gametophytes and sporophytes, showed diploid nuclei in the sporophytic filaments (Calderon-Saenz & Schnetter, 1989). Thus, in this species a heterokaryotic phase is lacking, and karyogamy obviously does not occur in its sporangia, but rather in the zygotes.

In the genera *Bryopsis* Lamouroux and *Pseudobryopsis* Berthold, the life cycle is composed of two generations. For a long time, the well-known macrothalli of both genera, which produce gametes, have been considered to be diploid. This opinion was mainly based on the publications of Schussnig (1932) and Zinnecker (1935) who observed gamete release, gamete fusion, and the germination of the zygotes in *Bryopsis*. They proposed that the germinating zygote develops into a new plumose *Bryopsis* thallus. Neumann (1969a, 1970) and Rietema (1969, 1975) were able to show that the germinating zygotes of *Bryopis* and *Pseudobryopsis* do not develop into plumose gametophytes, but into more or less irregularly forked microthalli. The microthalli can either develop directly into macrothalli, or the whole content of a macrothallus may be transformed into stephanokontan spores (Rietema, 1975). In the latter case, the zoospore germlings develop into new macrothalli. The same has been shown for *Pseudobryopsis* (Mayhoub, 1974; Kobara & Chihara, 1978a, b; Okuda et al., 1979; Wildner, unpublished). While most authors observed only one nucleus in the microthallus, which has been compared by Neumann (1969a, 1974) with the giant primary nucleus of *Acetabularia*, multinucleate young microthalli have been reported by Feldmann (1969) and Mayhoub (1974) for *Pseudobryopis*.

We observed that the zygote of *Pseudobryopsis oahuensis* Egerod in our cultures has only one nucleus, which is conserved in the growing microthallus during several months. Similar to the giant nucleus in *Acetabularia,* the nucleus of the microthallus is situated within a defined area. In regard, to the formerly zygote as the starting point of growth, the nucleus is always found near the first or second ramification of the microthallus.

There is no doubt that karyogamy takes place in *Bryopsis* and *Pseudobryopsis* after gamete fusion. As a result, the single nuclus in the microthalli, as observed by most authors, must be diploid. Evidence for the site of meiosis has not yet been reported, and contrary opinions concerning the ploidy level of the macrothalli exist in the literature. After the detection of the microthalli, most authors consider the macrothalli as haploid (e.g. Rietema, 1975). However, Kermarrec (1980) has postulated diploid macrothalli with gametic meiosis for *Bryopsis plumosa* (Hudson) C. Agardh and *B. hypnoides* Lamouroux. It may be concluded from a comparison of nuclear DNA contents in gametes and in the macrothallus of *Pseudobryopsis oahuensis* , that the macrothallus is haploid (Fig. 4): gametes and macrothallus filaments have identical nuclear DNA ranges. Probably the first division of the primary nucleus of the macrothallus is meiotic. A short transitional multinuclear stage must exist at least in the holocarpic microthalli of *Bryopsis* and *Pseudobryopsis,* releasing stephanokontan zoospores. This multinucleate stage may be heterokaryotic. It is interesting, that the production of anisogametes by microthalli in a strain of *Bryopsis hypnoides* from Newfoundland has been observed by Bartlett & South (1973). These microthalli developed later directly into macrothalli, which also produced anisogametes. Macrothalli of both genera, derived directly from microthalli, are also perhaps heterokaryotic. This opinion is supported by the observation of homothallic gametophytes (Feldmann, 1969; Kobara & Chihara, 1978a) and even hermaphroditic gametangia (Burr & West, 1970).

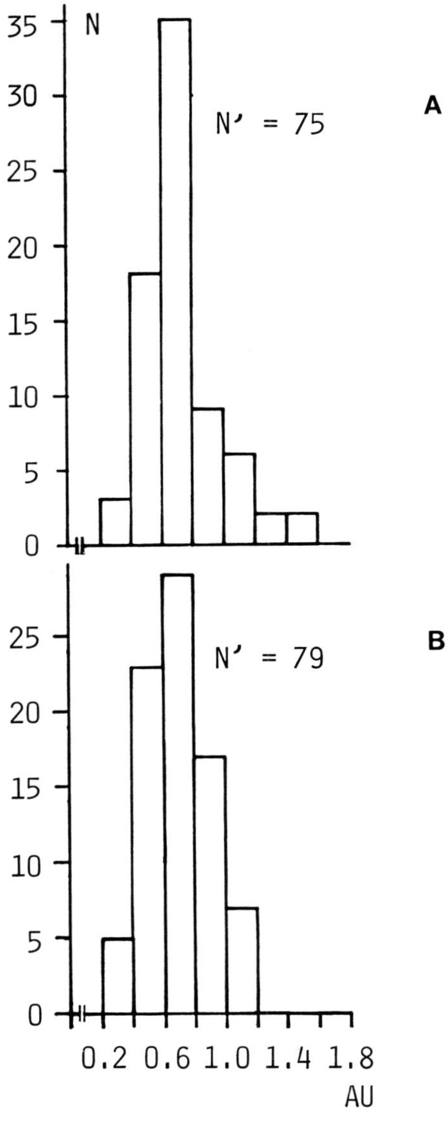

Fig. 4 : Relative nuclear DNA content in macrothalli (A) and Gametes (B) of *Pseudobryopsis oahuensis* , expressed in arbitrary units (AU). Ranges of DNA content show that the macrothallus nuclei are haploid. (Abbrevations and methods :see Fig. 1).

Little is known about the nuclei in the heteroplastidic Caulerpaceae and Udoteaceae. In *Caulerpa* Lamouroux (Schussnig, 1929; Price, 1972; Enomoto & Ohba, 1987) and *Udotea* Lamouroux (Meinesz, 1980), anisogametes fuse.

The zygotes develop into uninucleate protospheres, which are probably comparable with the microthalli of *Bryopsis* and *Pseudobryopsis*. Like the microthalli of several *Bryopsis* and *Pseudobryopsis* species, the protospheres germinate after several weeks or even months. Filamentous structures are formed out of which normal Caulerpa or *Udotea* thalli develop (Enomoto & Ohba, 1987). Schussnig (1939) reported meiotic nuclear divisions in adult thalli of *Caulerpa*. In general, the large *Caulerpa* thalli are therefore considered to be diploid (see also Goldstein & Morrall, 1970). Van den Hoek et al., (1988) recently supposed that the ploidy level changes in Bryopsidales with uninucleate protospheres (*Caulerpa, Halimeda* Lamouroux, *Udotea*) might be similar to those in *Acetabularia* and that the gamete producing plants should then be haploid.

CONCLUSIONS

It is evident that heterokaryotic life cycle phases are relatively widespread within the siphonous Chlorophyta. Two types of heterokaryotic phases may be distinguished: In *Acetabularia*, and probably in *Bryopsis* and *Pseudobryopsis*, the heterokaryon starts with the meiotic division of a diploid nucleus formed after gamete fusion. This heterokaryon is postsynkaryotic. In *Derbesia*, the heterokaryon starts with gamete mating and ends with karyogamy in the sporangia. This heterokaryon is hence presynkaryotic.

If *Bryopsidella ostreobiformis*, with its nearly isomorphic life cycle, is considered to be an evolutionary primitive organism, two different phylogenetic lines may then be distinguished within the Bryopsidaceae: the morphological reduction of the gametophyte (*Derbesia, ? Pedobesia*) is associated with the formation of a presynkaryotic heterokaryon, whereas the morphological reduction of the sporophyte (*Bryopsis, Pseudobryopsis*) is linked with a postsynkaryotic heterokaryon.

Further studies will show whether the uninucleate protospheres in the heteroplastidic genera *Udotea* and *Caulerpa* are comparable with the uninucleate microthalli of *Bryopsis* and *Pseudobryopsis*. In the latter two genera, microthalli exist with a holocarpic formation of stephanokontan spores as well as microthalli (of other species or strains) germinating directly into the gametophytic macrothalli. In this case, the gametes are the only motile stages, as in *Udotea* and *Caulerpa*. Recently detected stephanokontan spores in the heteroplastidic genus *Boodleopsis* indicate perhaps that primitive ancestors of *Udotea* and *Caulerpa* also produced stephanokontan zoospores in a life cycle phase corresponding to the protosphere in recent species.

The possible loss of zoospores in ancestors of *Caulerpa* during evolution has already been pointed out by Schussnig (1939). Our knowledge on the nuclear behaviour of siphonous Chlorophyta is still very fragmentary. The function of diploid and haploid thalli observed in species of the genus *Codium* Stackhause (Kapraun & Martin, 1987) within the life cycle is not completely understood, and further studies are necessary. Convincing data on ploidy level changes during life history in taxa of siphonous Chlorophyta not mentioned in this paper are still missing.

Financial support by a research grant from the Deutsche Forschungsgemeinschaft (Schn 88/6-1, 8-1) is greatfully acknowledged. The authors are indebted to Dr. J. Knappe, Marburg, FRG, for sending material of *Derbesia marina*.

The present paper in part contains results of a doctoral thesis of B.B.

REFERENCES

Abélard Ch (1982) A propos d'une nouvelle espèce de *Pedobesia* (Chlorophyeae, Derbesiales) provenant des Iles Galapagos: *Pedobesia feldmannii*. Cryptogam., Algol. 3: 187 - 209

Abélard Ch (1986) Mise au point sur la spécifité du *Pedobesia feldmannii* Abélard (Chlorophyceae, Derbesiales). Cryptogam., Algol. 7: 169 - 170

Abélard Ch, Knoepffler M (1986) *Pedobesia solieri* nov. sp. (Chlorophyta, Derbesiales) en Méditerranéen occidentale: morphologie et reproduction. Cryptogam., Algol. 7: 301 - 316

Al-Kubaisy KH, Schwantes HO, Seibold G (1981) Cytophotometrische Untersuchungen zum Generationswechsel autotropher und heterotropher siphonaler Organismen (*Vaucheria sessilis* und *Saprolegnia ferax*). Nova Hedwigia 34: ·301 - 316

Bartlett RB, South GR (1973) Observations on the life-history of *Bryopsis hypnoides* in culture. Acta Bot. Neerl. 22: 1 - 5

Bonotto S (1988) Recent progress in research on *Acetabularia* and related Dasycladales. Progr. Phycol. Res. 6: 159 - 235

Burr FA, West JA (1970) Light and electron microscope observations on the vegetative and reproductive structures of *Bryopsis hypnoides*. Phycologia 9: 17 - 37

Calderón-Sáenz E, Schnetter R (1989) Life cycle and morphology of *Bryopsidella ostreobiformis* spec. nov. (Bryopsidaecae, Chlorophyta) from the Mediterranean, under culture conditions, with comments on the phylogeny of the *Bryopsis/Derbesia* complex. Botanica Acta 102: 241 - 260

De Paula EJ, West JA (1986) Culture studies on *Pedobesia ryukyuensis* (Chlorophyta - Derbesiales), a new record in Brazil. Phycologia 25: 482 - 493

Eckhardt R, Schnetter R (1984) Failure of karyogamy after gamete mating in *Derbesia tenuissima*. Naturwissenschaften 71:

640 - 641
Eckhardt R, Schnetter R, Seibold G (1986) Nuclear behaviour during the life cycle of *Derbesia* (Chlorophyceae). Br. Phycol. J. 21: 287 - 295
Enoto S, Ohba H (1987) Culture studies on *Caulerpa* (Caulerpales, Chlorophycea) I. Reproduction and development of *C. racemosa* var. *laetevirens*. J. Jap. Bot. 35: 167 - 177
Feldmann J (1950) Sur l'existence d'une alternance de générations entre l'*Halicystis parvula* et le *Derbesia tenuissima* (De Not.) Crn.. C. r. hebd. Séance Acad. Sci. Paris, Sér. D 230: 322 - 323
Feldmann J (1969) *Pseudobryopsis myura* and its reproduction. Am. J. Bot. 56: 691- 695
Goldstein M, Morrall S (1970) Gametogenesis and fertilization in *Caulerpa*. Ann. N. Y. Acad. Sci. 175:660 - 672
Hoek C van den, Stamm WT, Olsen JL (1988) The emergence of a new chlorophytan system, and Dr. Kornmann's contribution thereto. Helgoländer Meeresunters. 42: 339 - 383
Kapraun DF, Martin DJ (1987) Karyological studies of three species of *Codium* (Codiales, Chlorophyta) from coastal North Carolina. Phycologia 26: 228 - 234
Kermarrec A (1980) Sur la place de la méiose dans le cycle de deux Chlorophycées marines: *Bryopsis plumosa* (Huds.) C. Ag. et *Bryopsis hypnoides* Lamouroux (Codiales). Cah. Biol.Mar. 21: 443 - 466
Kobara T, Chihara M (1978 a) On the taxonomy and sexual reproduction of the siphonous green alga *Pseudobryopsis hainanensis* Tseng. J. Jap. Bot. 53: 341 - 352
Kobara T, Chihara M (1978 b) On the life history of *Pseudobryopsis hainanensis* Tseng. J. Jap. Bot. 53: 353 - 360
Kobara T, Chihara M (1984) Laboratory culture and taxonomy of two species of *Pedobesia* (Bryopsidales, Chlorophyceae) in Japan. Bot. Mag. Tokyo 97: 151 - 161
Kornmann P (1938) Zur Entwicklungsgeschichte von *Derbesia* und *Halicystis*. Planta 28: 464 - 470
Liddle L, Berger S, Schweiger HG (1976) Ultrastructure during development of the nucleus of *Batophora oerstedii* (Chlorophyta, Dasycladales). J. Phycol. 12: 261 - 272
MacRaild GN, Womersley HBS (1974) The morphology and reproduction of *Derbesia clavaeformis* (J. Agardh) De Toni (Chlorophyta). Phycologia 13: 83 - 93
Mayhoub H (1974) Reproduction sexuée et cycle du dévelopment de *Pseudobryopsis myura* (Ag.) Berthold (Chlorophycée, Codiale). C. R. Acad. Sci. Paris 278, sér. D: 867 - 870
Meinesz A (1980) Connaissances actuelles et contribution à l'étude de la reproduction et du cycle des Udotéacées (Caulerpales, Chlorophytes). Phycologia 19: 110 - 138
Menzel D (1980) Plug formation and peroxidase accumulation in two orders of siphonous green algae (Caulerpales and Dasycladales) in relation to fertilization and injury. Phycologia 19: 37 - 48
Neumann K (1967) Der Ort der Meiosis und die Sporenbildung bei der siphonalen Grünalge *Derbesia marina*. Naturwissenschaften 54: 121
Neumann K (1969 a) Protonoma mit Riesenkern bei der siphonalen Grünalge *Bryopsis hypnoides* und weitere cytologische Befunde. Helgoländer wiss. Meeresunters. 19: 45 - 87

Neumann K (1969 b) Beitrag zur Cytologie der siphonalen Grünalge *Derbesia marina*. Helgoländer wiss. Meeresunters. 19: 355 - 375

Neumann K (1970) Einkerniges Protonema bei *Bryopsis* and *Pseudobryopsis myura*. Helgoländer wiss. Meeresunters. 20: 213 - 215

Neumann K (1974) Zur Entwicklungsgeschichte und Systematik der siphonalen Grünalgen *Derbesia* und *Bryopsis*. Bot. Mar. 17: 185

Okuda K, Enomoto S, Tatewaki M (1979) Life history of *Pseudobryopsis spec*. (Codiales, Chlorophyta). Jap. J. Phycol. 27: 7 - 27

Price IR (1972) Zygote development in *Caulerpa* (Chlorophyta, Caulerpales). Phycologia 11: 217

Rietema H (1969) A new type of life history in *Bryopsis* (Chlorophyceae, Caulerpales). Acta bot. Neerl. 18: 615 - 619

Rietema H (1972) A morphological, developmental, and caryological study on the life history of *Bryopsis halymeniae* (Chlorophyceae). Neth. J. Sea Res. 5: 445 - 457

Rietema H (1973) The influence of day length on the morphology of the *Halicystis parvula* phase of *Derbesia tenuissima* (De Not.) Crn. (Chlorophyceae, Caulerpales). Phycologia 12: 11 - 16

Rietema H (1975) Comparative investigations on the life histories and reproduction of some species in the siphoneous green algal genera *Bryopsis* and *Derbesia*. Proefschrift, Groningen: 130 pp.

Schnetter R, Mohr B, Bula-Meyer G, Seibold G (1981) Ecology, life history and nucleus DNA contents of *Derbesia tenuissima* from the Caribbean coast of Colombia. Xth International Seaweed Symposium (Levring, T., ed.): 357 - 362

Schnetter R, Eckhardt R, Seibold G (1985) Die Kernphase von Sporophytenfäden der Grünalge *Derbesia tenuissima* (De Not.) Crn.. Beitr. Biol. Pflanzen 60: 293 - 302

Schussnig B (1929) Die Fortpflanzung von *Caulerpa prolifera*. Öster. Bot. Zschr. 78: 1 - 8

Schussnig B (1932) Der Generations- und Phasenwechsel bei den Chlorophyceen. III. Beitrag. Öster.Bot. J. 81. 296 - 298

Schussnig B (1939) Ein Beitrag zur Entwicklungsgeschichte von *Caulerpa prolifera*. Bot. Notizen 1939: 75 - 96

Tanner CE (1981) Chlorophyta: Life histories. In: Lobban, C.S. and Wynne, M.J., eds., - The Biology of Seaweeds (Bot. Monogr. 17), Blackwell Sc. Publ., Oxford: 218 - 247

Zinnecker H (1935) Beiträge zur Entwicklungsgeschichte der Protophyten. XII. Reduktionsteilung, Kernphasenwechsel und Geschlechtsbestimmung bei *Bryopsis plumosa* (Huds.) Ag.. Öster. Bot. Z. 84: 53 - 72

The polyspermy block in fucoid algae

Susan H. Brawley
Department of Biology
Vanderbilt University
Nashville TN 37235
USA

Introduction

It is as important for an egg to be fertilized by only one sperm as to be fertilized at all. "Polyspermy" is the condition of an egg fertilized by more than one sperm, and it is lethal in many species including fucoid algae, sea urchins, frogs, and man (Jaffe & Gould, 1985; Brawley, 1987; Kola & Trounson, 1989). Polyspermic zygotes die, not because of extra genetic material per se, but due to the maldistribution of chromosomes caused by extra centrioles, paternally-inherited, which create a multi-polar spindle. Some organisms (e.g., birds, reptiles and many urodeles) are "physiologically polyspermic" (Wilson, 1925; Iwao, 1989). Eggs of these species sustain multiple fertilizations, but only one set of sperm organelles becomes functional; the others are destroyed within the egg cytoplasm by as yet unknown means.

In a recent study (Brawley, 1987), I demonstrated that the monoecious fucoid algae Pelvetia fastigiata and Fucus distichus have a sodium-dependent, fast block against polyspermy. The degree to which Fucus distichus and Pelvetia fastigiata became polyspermic when fertilized in a low-sodium artificial seawater (2.5 mM Na^+ ASW) was different, however: 15% vs. 90% of eggs, respectively (Brawley, 1987). The present report seeks to explain this difference and presents related information on F. vesiculosus, a dioecious fucoid alga. Use of F. vesiculosus allows the timing of the fast block and the effect of various sperm-egg ratios on polyspermy to be tested more critically than is possible with monoecious species. Finally, information on the electrophysiological changes at fertilization is presented, showing why sodium limitation increases polyspermy, and I speculate on why eggs in fucoid algae, under permissive conditions for polyspermy, still allow only a few extra sperm to enter.

Materials and Methods

Collection and Culture. Fucus vesiculosus was collected at Nahant and Manchester, MA (USA) between October and January, 1988-89; Fucus distichus was collected at Pacific Grove, CA (USA). Gametes were obtained from receptacles stored for no more than 3 d. The best release of F. vesiculosus gametes was obtained as follows. Receptacles were rinsed in cold (10°C) tap water for 2 min (males) or for 5-7 min (females). Receptacles were patted dry between paper towelling. Female receptacles were covered with seawater in glass baking dishes, placed at 10-15°C in bright light (typically outdoors) for 1-2 h and then transferred to a dimly-lit culture chamber (65 μE/m ^{2}/sec); massive shedding of oogonia occurred within 10 min. Antheridia were shed from ripe male receptacles that had been rinsed briefly in seawater and then exposed to air, outdoors (4-15°C), for 30 min - 1 h. Shedding appears to be a response to desiccation and is observed in nature, especially as receptacles dry out on cold, autumn days in the intertidal zone.

Procedures for release of gametes from other species, culture of zygotes, and preparation of artificial seawaters were as given by Brawley (1987). The sodium substitute used in these experiments was n-methyl glucamine.

Effect of sperm-egg ratios on polyspermy. Suspensions of F. vesiculosus sperm were obtained by dipping receptacles covered with freshly-shed antheridia into a small quantity of seawater. Debris was removed from the suspension with 30 sec of centrifugation at speed 3 (1000 rpm) using an International Clinical centrifuge (Needham, MA). The sperm suspension was gently mixed with a pipette, and the sperm concentration determined by fixing an aliquot in 1% paraformaldehyde in seawater and counting in a Levy-Hausser chamber (Hausser Sci., Blue Bell, PA). The suspension was diluted as necessary and added to eggs at the desired ratio. Egg concentration was determined volumetrically; eggs were not centrifuged but were allowed to settle by gravity in centrifuge tubes (3.5×10^4 eggs/100 μl). Gametes were used within 15 min of shedding from oogonia or antheridia, and they were mixed in glass petri dishes (60 mm diameter) containing normal or artificial seawaters (pH 7.9, 15°C, 5 ml). Polyspermy was determined both by direct staining of fixed zygotes in aceto-iron-hematoxylin and, indirectly, by the distinctive morphology of polyspermic fucoid embryos (Brawley, 1987; Fig. 1, this report).

Time requirement for fast block. Gametes were mixed (sperm:egg, 800:1) in normal seawater (pH 7.9, 15°C) in a small beaker at t=0 min. At 1 to 5-min intervals thereafter, eggs were transferred into 45 mM Na^+ ASW (n-methyl glucamine substituted) containing fresh sperm (800:1). Eggs (and some sperm) were transferred quantitatively by letting eggs settle in volumetric centrifuge tubes over ca. 20 sec before transfer. After 60 min, zygotes were fixed in acetic alcohol (1:3) and stained (Brawley, 1987) to determine polyspermy. The time of fertilization after mixing of eggs and sperm was determined by transferring a portion of eggs into 1% chloral hydrate (a spermicide) in seawater at 1-min intervals after t=0, at the same time as transfers of eggs were made to low-sodium ASW. After 1 h, eggs were assayed for fertilization using the Calcofluor White assay for cell wall formation (Callow et al., 1978; Brawley & Bell, 1987).

Electrophysiology. Eggs were impaled with 70-90 MΩ glass electrodes (1.2 mm O.D. glass capillary, W.P.I., Inc., New Haven, CT) pulled with a horizontal electrode puller (ISA, Inc., Flushing, NY). They were back-filled with 1.5 M potassium acetate (pH 7.0 with HCl). The eggs were immobilized by attachment to cover slips coated with poly-lysine (cover slips dipped in poly-D-lysine hydrobromide, 150-300 kD, at 2 mg/ml in distilled water for 10 sec). Membrane potential was recorded using a Biodyne (Santa Monica, CA) AM-2 voltage follower, Gould (Cleveland, OH) 220 chart recorder and Nicolet 310 oscilloscope. The bath was grounded through a seawater-2% agar bridge, and the bath was maintained at ambient seawater temperature by placing the petri dish in a seawater-cooled cold stage on the Zeiss inverted microscope. If the electrode's tip potential changed by more than 3 mV, data were not used.

Sodium-selective electrodes. The free Na^+ content of F. distichus mucilage was determined using a Na^+-selective mini-electrode (Microelectrodes, Inc., Londonderry, NH). A standard curve of 0.1 mM - 1M NaCl and 1M KCl was used to calibrate the electrode per the manufacturer's instructions. Mucilage samples were collected from 10 different plants along a transect through the F. distichus zone by collecting mucilage from the surface of a single receptacle 2-3 h after plants were exposed at low tide; this was the time at which oogonia and antheridia were being shed from receptacles. Simultaneously, from another receptacle on each plant, gametes (+ mucilage) were pipetted into 350 mM K^+ ASW, which prevents fertilization (Robinson et al., 1981); these eggs were assayed for fertilization with Calcofluor White (see above) in the laboratory. Mucilage samples were determined directly as a 100 μl droplet and by dilution of 10 μl of mucilage in 90 μl of 1 mM Na^+ solution.

Results

Polyspermic embryos of F. vesiculosus (Fig. 1) have the stunted morphology that is typical of polyspermy in P. fastigiata and F. distichus (Brawley, 1987). Most of these embryos die by 3 d post-fertilization, and all are dead by 7 d post-fertilization. Polyspermy is induced by both higher sperm-egg ratios and fertilization of eggs in low-sodium seawater (Fig. 2). About 53% of embryos are polyspermic when fertilized in 10 mM Na^+ ASW at a sperm-egg ratio of 2100:1. When eggs are fertilized under conditions that are increasingly permissive for polyspermy, there is an increase not only in the total number of polyspermic eggs but in the number of sperm penetrating each egg. At a sperm-egg ratio of 2100:1, for example, eggs fertilized in 10 mM Na^+ ASW sometimes contain 5 sperm pronuclei, and as many eggs are trispermic as dispermic or monospermic (Table 1).

Sodium ions are required for only a short time after fertilization to block polyspermy in F. vesiculosus. At an initial sperm:egg ratio of 800:1, 50% of the eggs were fertilized by 1.5 min after mixture (at t=0 min) of gametes, and there was a 50% decrease in polyspermy when eggs were transferred to low-Na^+ ASW

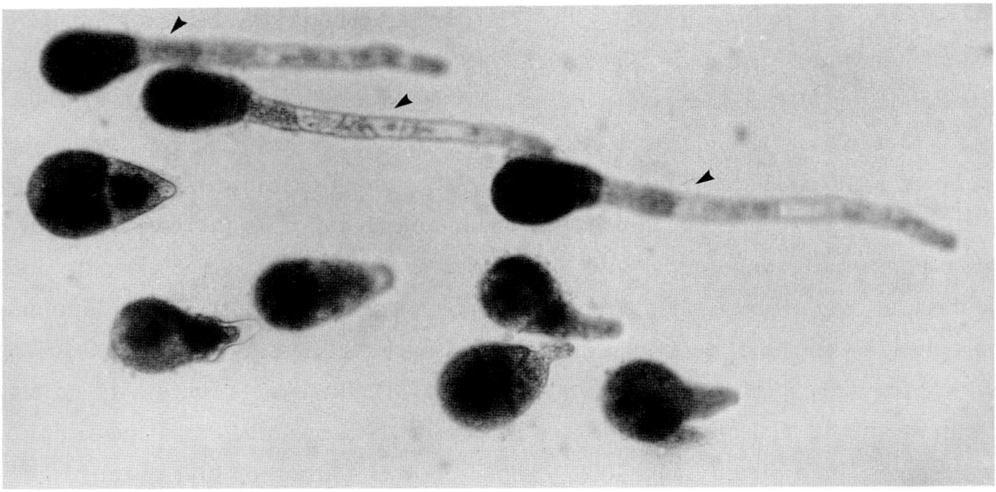

Figure 1. Polyspermic and normal (arrows) embryos of Fucus vesiculosus at 7 d post-fertilization.

at t=3 min. This suggests that Na^+ ions are required for 1.5 - 2 min post-fertilization (Fig. 3) to block polyspermy. Supporting this conclusion is an experiment using a higher initial sperm-egg ratio (2300:1), which likely increased the speed of fertilization (not assayed); polyspermy was not evident above control values in transfers done later than 2 min after gametes were mixed.

The reason that sodium ions are required to block polyspermy is suggested by observing the egg's membrane potential during fertilization in 452 mM Na^+ ASW (normal ASW) vs. 45 mM Na^+ ASW (Fig. 4). The fertilization potential is 1) shorter, 2) less positive-going, and 3) reaches its depolarized plateau more slowly in low-sodium ASW than in normal ASW. A full analysis of the fucoid fertilization potential and its significance will be presented elsewhere (Brawley, in prep.), but the timing experiments presented here strongly suggest that the duration of the fertilization potential in low Na^+ (45 mM Na^+ ASW) is adequate to prevent polyspermy. The higher sensitivity of such eggs to polyspermy is, thus, likely related to either the slower depolarization following fertilization or the fact that the fertilization potential is about 10 mV more negative in 45 mM Na^+ ASW (Fig. 4;

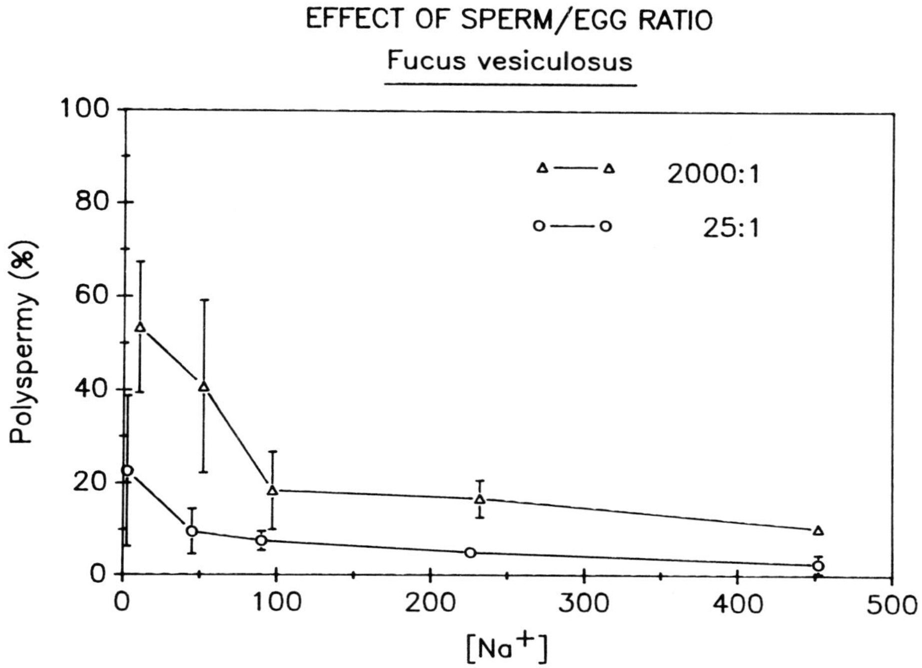

Figure 2. Polyspermy increases during fertilization at higher sperm-egg ratios and as more of the Na^+ in an artificial seawater is replaced by an impermeant cation. Values are the means (±S.D.) of 3 experiments.

Brawley, in prep.). Wall formation is triggered by fertilization in the Fucales and is first observed about 10 min post-fertilization in F. vesiculosus with the fluorescent dye Calcofluor White, as reported by Brawley & Bell (1987). It is likely that this is the slow, permanent block against polyspermy, and that it becomes effective prior to the time that wall material can be documented with light (fluorescence) microscopy. However, intermediate blocks (e.g., see Lambert, 1986, and Bolwell et al., 1980) against polyspermy may also be present in the fucoid egg.

Some members of the Fucales shed copious mucilage along with oogonia and/or antheridia when gamete release is stimulated. For example, this is true of

Table 1

[Na+] ASW	Sperm-Egg Ratio	Number of sperm pronuclei in egg					Polyspermy (%)
		1	2	3	4	5	
452 mM Na^+	25:1	98	1	1			2%
47 mM Na^+	25:1	95	5				5%
2 mM Na^+	25:1	70	25	4	1		30%
452 mM Na^+	250:1	94	5	1			6%
47 mM Na^+	250:1	66	28	6			44%
5 mM Na^+	250:1	46	29	19	6		64%
452 mM Na^+	2100:1	90	10				10%
47 mM Na^+	2100:1	34	43	20	2	1	66%
10 mM Na^+	2100:1	26	27	28	12	7	74%

Table 1. Number of sperm pronuclei in F. vesiculosus eggs (n = 100 eggs counted/treatment) as a function of external [Na^+] and sperm-egg ratio during fertilization.

the monoecious fucoid Fucus distichus but not of Pelvetia fastigiata, also monoecious. To determine whether mucilage is associated with Na^+ enrichment of low-Na^+ ASW (i.e., unexpected protection against polyspermy) in the experiments of Brawley (1987), the free Na^+ levels of mucilage were determined at the time shedding began in the field (see Materials and Methods). On two different days, mucilage was found to contain high levels of free Na^+ (Table 2). Many of the eggs shed during this low-tide period were fertilized shortly after release of oogonia and antheridia from the receptacle (i.e., before seawater recovered plants). Polyspermy was low, presumably because of the ionic (Na^+-rich) nature of the mucilage.

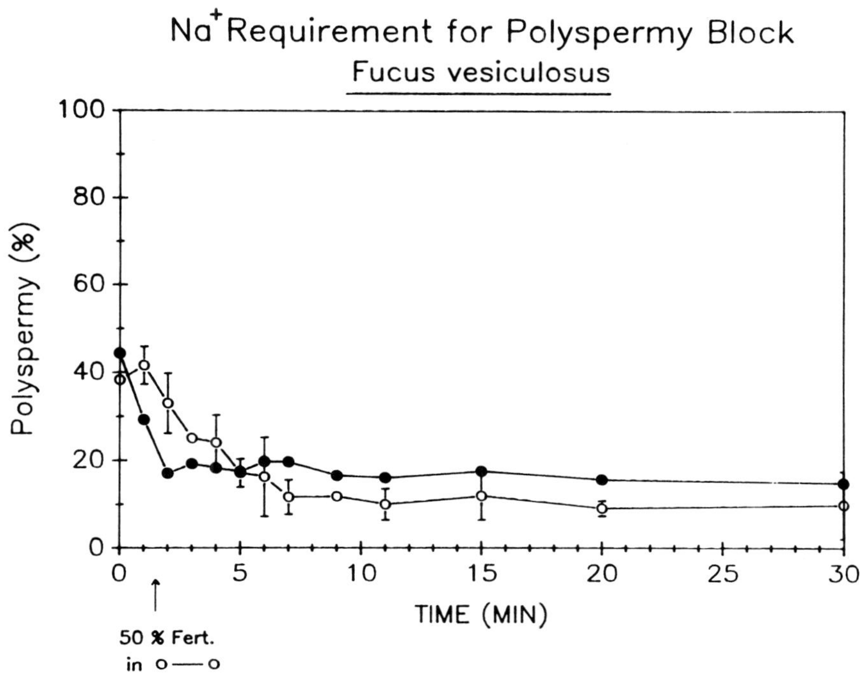

Figure 3. Sodium is required for only 1.5 - 2 min post-fertilization to block polyspermy. Eggs were fertilized in normal seawater and transferred to low-Na^+ ASW at the times shown on the abscissa. Experiment 1 (dark circles) was conducted at a higher sperm-egg ratio (see text) than those of the data represented by open circles (mean ± S.D. of 2 experiments).

Discussion

This report demonstrates that Fucus vesiculosus has a fast block against polyspermy, and it suggests that the sodium-dependence of the block is related to the deleterious effect of sodium limitation on the fertilization potential, which is described in low-sodium seawater for the first time. The report, also, suggests that the lower sensitivity to polyspermy in low-Na^+ ASW of Fucus distichus compared to Pelvetia fastigiata (Brawley, 1987) is because of sodium enrichment related to mucilage secretion. Indeed, in a solution that was initially 2.5 mM Na^+ ASW, free [Na^+] of 100 mM has been measured at the end of a F. distichus experiment conducted per Brawley (1987) (Brawley & Hassell, in prep.). However, it appears that Pelvetia eggs may be genuinely more sensitive to polyspermy than Fucus eggs, since the sperm-egg ratio in P. fastigiata at fertilization in the laboratory could not be more than 200:1 (Brawley, 1987), but polyspermy is much higher under equivalent conditions (2-5 mM Na^+ ASW) compared to F. vesiculosus at sperm-

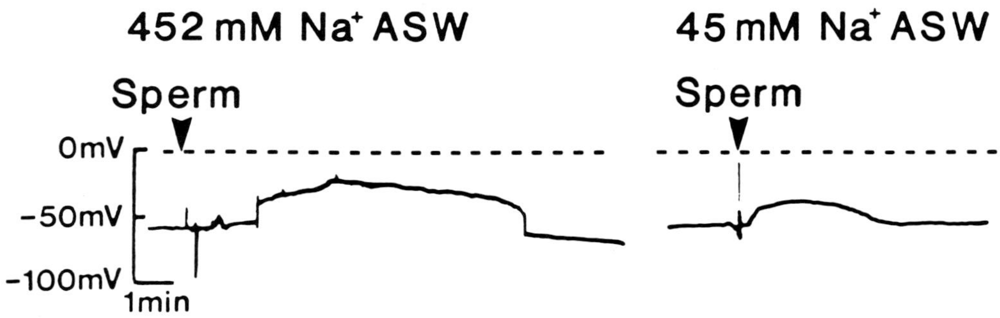

Figure 4. Fertilization potentials of Fucus vesiculosus in normal and low-sodium artificial seawater.

egg ratios of 250:1 (Brawley, unpublished) or 2000:1 (see Results). This might be related to the larger size of the Pelvetia egg. Also, the "sperm pressure" may be higher in Pelvetia than in either species of Fucus studied thus far; as noted in Brawley (1987), antheridia are often shed attached to the oogonium of Pelvetia, and this has not been observed in Fucus vesiculosus or F. distichus.

In some animal eggs, dozens of additional sperm penetrate an egg fertilized in low-sodium ASW (e.g., Urechis, Gould-Somero et al., 1979). Why is this not true of fucoid eggs? One possibility is suggested by the behaviour of sperm at the egg surface. Sperm rarely fuse with the egg immediately (i.e., elicit a fertilization potential). Instead, they "search" the surface of the egg for up to 30 sec. This suggests that the numbers of some receptor molecule(s) may be limiting, although

Table 2

	5/3/89	5/16/89
$[Na^+]$ in mucilage (n = 10 receptacles)	580 (151) mM	460 (100) mM
Fertilization during low tide (% of 250 eggs) (n = 4 receptacles, 5/3; 5 receptacles, 5/16)	94 (7) %	81 (23) %
Polyspermy (% of 250 eggs) (n = 4 receptacles, 5/3; 5 receptacles, 5/16)	2.2 (1.0) %	2.4 (1.0) %

Table 2. Characteristics of mucilage shed with gametes from receptacles of *Fucus distichus*. Values are means (±S.D.).

there is no reason at present to suspect that localization of receptor molecules occurs.

Polyspermy has been observed in algae other than the fucoid algae (e.g., *Oedogonium*, Hoffman, 1973); however, how widespread the lethal character of polyspermy is remains to be determined for other algae and plants. An important question, in this regard, is whether centriole or centrosome inheritance occurs through the egg or sperm; this is unknown in most plants. The *Oedogonium* sperm, which has multiple basal bodies, and organisms such as the red algae and angiosperms, which have centrosomes but no centrioles, will be of special interest. There are many algae in which high sperm-egg ratios may occur at fertilization because of the presence of sexual pheromones (see Maier & Müller, 1986). Therefore, the eggs of many species may require a polyspermy block for normal development.

Acknowledgements

Special thanks to David Epel and other faculty and staff of Hopkins Marine Station (Stanford University) and to the faculty and staff of the Marine Science Center (Nahant, MA) of Northeastern University for their hospitality during a sabbatical leave from Vanderbilt University, 1988-89. Michael Claes was a loyal collector, Grace Monty provided word-processing, and this work was supported by the National Science Foundation (U.S.A.).

References

Bolwell GP, Callow JA, Evans LV (1980) Fertilization in brown algae III. Preliminary characterization of putative gamete receptors from eggs and sperm of Fucus serratus. J Cell Sci 43: 209-224

Brawley SH (1987) A sodium-dependent, fast block to polyspermy occurs in eggs of fucoid algae. Dev Biol 124: 390-397

Brawley SH, Bell E (1987) Partial activation of Fucus eggs with calcium ionophores and low-sodium seawater. Dev Biol 122: 217-226

Callow ME, Evans LV, Bolwell GP, Callow JA (1978) Fertilization in brown algae I. SEM and other observations on Fucus serratus. J Cell Sci 32: 45-54

Gould-Somero M, Jaffe LA, Holland LZ (1979) Electrically-mediated fast polyspermy block of the marine worm, Urechis caupo. J Cell Biol 82: 426-440

Hoffman LR (1973) Fertilization in Oedogonium II. Polyspermy. J Phycol 9: 296-301

Iwao Y (1989) An electrically-mediated block to polyspermy in the primitive urodele Hynobius nebulosus and phylogenetic comparison with other amphibians. Dev Biol 134: 435-445

Jaffe LA, Gould M (1985) Polyspermy-preventing mechanisms. In Metz CB, Monroy A (eds) "Biology of Fertilization" Vol 3. Academic Press, New York, pp 223-250

Kola I, Trounson A (1989) Dispermic human fertilization: violation of expected cell behavior. In: Schatten H, Schatten G (eds) The Cell Biology of Fertilization Academic Press, San Diego, pp 277-293

Lambert CC (1986) Fertilization-induced modification of chorion N-acetyl-glucosamine groups in the ascidian chorion, decreasing sperm binding and polyspermy. Dev Biol 116: 168-173

Maier I, Müller DG (1986) Sexual pheromones in algae. Biol Bull 170: 145-75

Robinson KR, Jaffe LA, Brawley SH (1981) Electrophysiological properties of fucoid algal eggs during fertilization. J Cell Biol 91: 179a

Wilson EB (1928) The Cell in Development and Heredity, 3rd edn. Macmillan Co, New York

The evolution of algal chloroplasts

Sarah P. Gibbs
Department of Biology
McGill University
1205 Avenue Docteur Penfield
Montreal, Quebec H3A 1B1
Canada

The algae are not a natural assemblage of organisms. Rather they are a diverse group of protists and fungi which have acquired chloroplasts in various ways, in some groups directly from symbiotic photosynthetic prokaryotes and in others from symbiotic eukaryotic algae. The new techniques for rapidly sequencing ribosomal RNA, discussed by Adoutte in this volume, are producing evolutionary trees which are giving us a clearer understanding of the true relationships between different eukaryotes. In Fig. 1, I have drawn an evolutionary tree based on the sequences of the small subunit ribosomal RNA of a number of eukaryotes. The most ancient eukaryotes whose ribosomal RNA has been sequenced to date are *Giardia lamblia*, a parasitic diplomonad, and *Vairimorpha necatrix*, a microsporidian. Neither of these protists has mitochondria, and it is possible that these ancient eukaryotes evolved prior to the acquisition of mitochondria. Also ancient are the trypanosomes and *Euglena*. Somewhat later *Dictyostelium* evolved, but then came an explosive radiation of many protist groups, fungi, plants and animals. In this tree, I have put in boldface lettering those groups which contain species with chloroplasts. A glance at the tree shows that the algae are polyphyletic. Euglenoids evolved very early and are related to trypanosomes. The dinoflagellates, which many have assumed to be an ancient group because of the presence of a mesokaryotic nucleus (Loeblich, 1976), in fact evolved late and are related to ciliates. The brown algae and Chrysophytes are related to each other and to the Oomycetes, a relationship long suspected due to the presence of heterokont flagella in each group. Higher plants and green algae form a natural assemblage. The fact that organisms with chloroplasts appear on different branches of the tree indicates that different groups have acquired chloroplasts independently of each other. In this symposium presentation, I will summarize our present understanding of how chloroplasts evolved in the different groups of algae.

In only two groups of algae, the green algae and the red algae, am I certain that chloroplasts evolved directly from symbiotic prokaryotic cells. These are the only algae in which the chloroplasts are surrounded by two membranes. The inner of these membranes is postulated to have evolved from the plasmalemma of the prokaryotic cell and the outer from the phagocytic vacuole membrane of the eukaryotic host. There is no doubt that the chloroplasts of red algae have evolved directly from symbiotic cyanobacteria (Bonen and Doolittle, 1975). The striking similarity between red algal chloroplasts and cyanobacteria, with each possessing prominent phycobilisomes on the outer surfaces of their thylakoids, attest to this fact. Whether green algae evolved from a red algal ancestor (the monophyletic origin of chloroplasts, Cavalier-Smith, 1982) or whether green algae independently acquired chloroplasts either from symbiotic cyanobacteria or from symbiotic prochlorophytes (the polyphyletic origin of chloroplasts, Raven, 1970) is still debated. Most of the debate centers around whether green algal chloroplasts are more similar to cyanobacteria or prochlorophytes. In ultrastructure and pigment composition, green algal chloroplasts are strikingly similar to *Prochloron* and *Prochlorothrix*. However, the sequence of the 16S ribosomal RNA from the chloroplast of *Chlamydomonas reinhardtii* is closer to cyanobacterial 16S ribosomal RNAs than to *Prochlorothrix* 16S ribosomal RNA (Turner et al., 1989). In contrast, a parsimony tree based on the sequences of the *psbA* gene places green chloroplasts closer to *Prochlorothrix* than to cyanobacteria (Morden and Golden, 1989). At present, it is not possible to know whether green algae acquired their chloroplasts from symbiotic cyanobacteria or symbiotic prochlorophytes. However, there is considerable evidence, summarized by Gray and Doolittle (1982), which suggests that green algae acquired their chloroplasts from prokaryotes through a different symbiosis from that which gave rise to the chloroplasts of red algae. In addition, the many morphological differences between red and green algae, the differences between red algal cytoplasmic 5S ribosomal RNA and that of other eukaryotes (Van den Eynde et al., 1988) and the distance between red and green algae in an evolutionary tree based on partial sequences of the large subunit ribosomal RNA (Perasso et al., 1989) further support the hypothesis that evolution of chloroplasts from prokaryotes occurred independently in the red and green algae.

Fig. 1 shows clearly that the euglenoids evolved long before any of the other chloroplast-containing organisms. There is no doubt that the euglenoids have acquired their chloroplasts independently. *Euglena* chloroplasts,

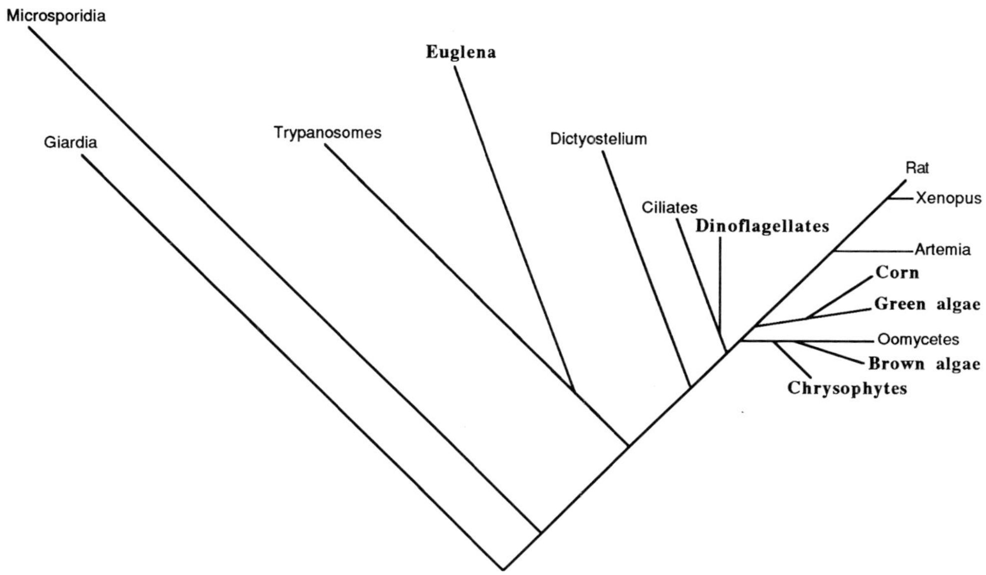

Fig. 1. Eukaryotic phylogenetic tree based on small subunit ribosomal RNA sequence comparisons. Chloroplast-containing species are present in the groups of organisms in boldface type. Tree redrawn from Fig. 14 in Woese (1987) with additions from Sogin et al. (1986), Gunderson et al. (1987), Bhattacharya and Druehl (1988) and Sogin et al. (1989)

along with those of dinoflagellates, are unique in that they are surrounded by three membranes. The presence of this third membrane was the clue that led me to postulate that euglenoids had acquired their chloroplasts from symbiotic green algae (Gibbs, 1978). I proposed that a colorless phagotrophic euglenoid engulfed a green alga and over time reduced it to a chloroplast surrounded by the persisting plasmalemma of the green alga. Although I still believe this is a likely scenario, it is possible that euglenoids engulfed either cyanobacteria or prochlorophytes and reduced them to chloroplasts, only in the evolution of Euglena chloroplasts, three membranes remained around the chloroplast, the inner and outer cell membrane of the Gram-negative prokaryote plus the phagocytic vacuole membrane of the euglenoid host cell. However, in a dendrogram based on the sequences of the 16S ribosomal RNAs of plastids and prokaryotes, the chloroplast of Euglena gracilis is on the same branch of the tree as the chloroplast of Chlamydomonas reinhardtii and is removed from cyanobacteria and even more distant from Prochlorothrix (Turner et al., 1989). This supports my belief that *Euglena* chloroplasts have evolved from green algae rather than cyanobacteria. However, in this dendrogram, the branches leading to *Euglena* and *Chlamydomonas* chloroplasts are long and in a tree with long branches and short stems, the branching order is more likely to be incorrect.

Although uncertainty exists as to the manner of evolution of chloroplasts surrounded by three membranes, I am certain that in all the classes of algae with chloroplasts surrounded by four membranes, the chloroplasts have evolved from an endosymbiotic eukaryotic alga. In eight classes of algae, namely the Cryptophyceae, Prymnesiophyceae, Raphidophyceae, Chrysophyceae, Bacillariophyceae, Phaeophyceae, Xanthophyceae and Eustigmatophyceae, the chloroplast is completely enclosed by an additional pair of membranes, named the chloroplast endoplasmic reticulum (Gibbs, 1981). Ribosomes are present on the cytoplasmic surface of the outermost membrane. Also, in many species, the outer membrane of the chloroplast ER is continuous with the outer membrane of the nuclear envelope in such a manner that a piece of the nuclear envelope forms part of the outer double membrane enclosing the chloroplast. This distinctive arrangement of a chloroplast surrounded by four membranes is, I believe, always the result of a cell having acquired a chloroplast from a symbiotic eukaryotic alga with the third membrane having evolved from the plasmalemma of the eukaryotic alga and the fourth membrane from the phagocytic vacuole membrane of the host cell. In seven of the eight classes, the chloroplast ER lies close to the chloroplast envelope and the only remnants of the endosymbiont's cytoplasm are the vesicles and tubules of the periplastidal reticulum (Gibbs, 1981), but in one class, the Cryptophyceae, the periplastidal space is much enlarged on the inner face of the chloroplast and contains both remnants of the endosymbiont's cytoplasm and a small vestigial nucleus named a nucleomorph (Greenwood et al., 1977). Fig. 2 shows the ultrastructure of the chloroplast and periplastidal compartment in *Cryptomonas abbreviata*. The enlarged periplastidal compartment on the inner face of the chloroplast contains starch grains, scattered vesicles and tubules, and clusters of ribosomes (arrows). These ribosomes are the same size as the eukaryotic ribosomes in the cryptomonad's cytoplasm and are distinctly larger than the chloroplast ribosomes. A prominent nucleomorph is also present in the periplastidal compartment lying in a groove on the inner face of the pyrenoid. The nucleomorph is limited by a double membrane which is interrupted by pores (not shown), which in some cryptomonad species are circular (Morrall and Greenwood, 1982) and in others elliptical (Gillott and Gibbs, 1980). The

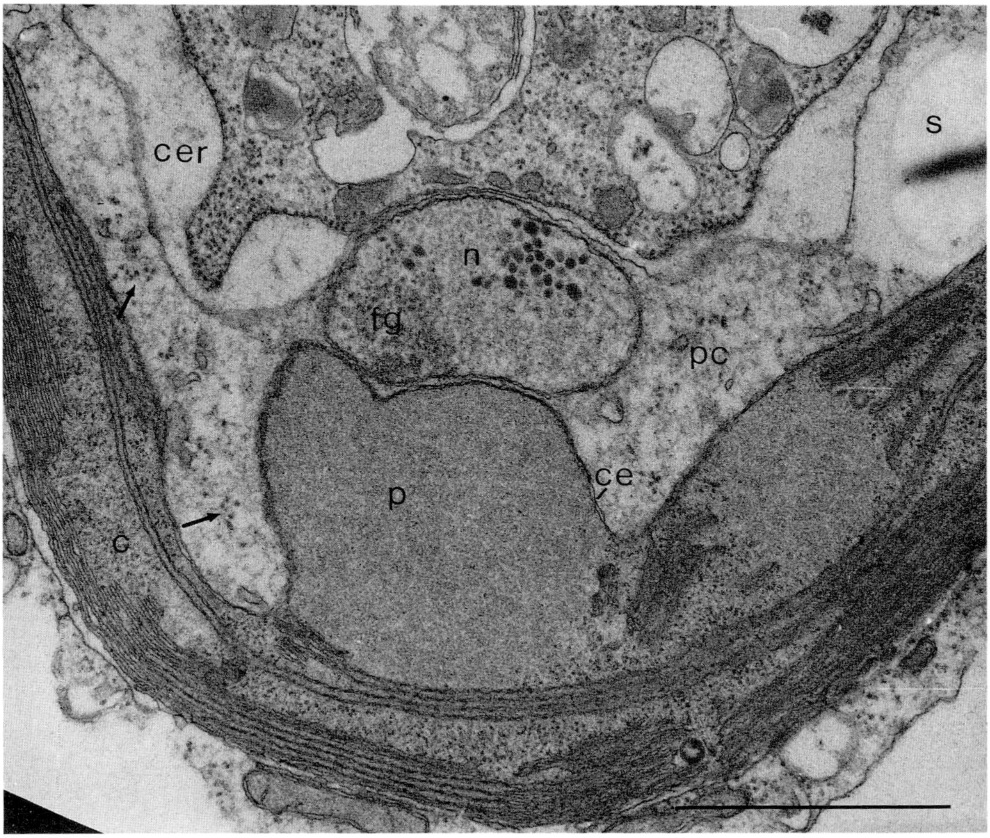

Fig. 2. Cross-section through the chloroplast (c) of Crytomonas abbreviata. The chloroplast ER (cer) delimits a wide periplastidal compartment (pc) on the inner face of the chloroplast. Starch grains (s), eukaryotic-sized ribosomes (arrows) and a nucleomorph (n) are located in the periplastidal compartment. The nucleomorph is enclosed by a double membrane and contains rod-shaped particles and a fibrillogranular body (fg). p, pyrenoid; ce, chloroplast envelope. Scale bar = 1.0 μm

nucleomorph contains a small nucleolus-like body, named the fibrillogranular body, which Bernhard's differential staining technique suggested contained RNA (Gillott and Gibbs, 1980), a result recently confirmed by Hansmann (1988) by labelling with RNase complexed to colloidal gold. The nature of the prominent dense globules (which have the shape of short rods and are sometimes hexagonally packed) is unknown, but they may be viruses. The nucleomorph divides by an infolding of the inner membrane followed by constriction in two just prior to prophase (McKerracher and Gibbs, 1982). The presence of DNA in the nucleomorph of two different species of cryptomonads has been demonstrated by DAPI fluorescence (Ludwig and Gibbs, 1985; Hansmann et al., 1985). After Ryter and Kellenberger's fixation, the matrix of the nucleomorph is seen to be full of naked DNA fibrils (Ludwig and Gibbs, 1985). We failed to observe labelling of the nucleomorph with antibodies to histone H2A and H3 of chicken erythrocytes. Hansmann et al. (1987) also obtained negative results with antibodies to histones H3 and H4. The presence of DNA in cryptomonad nucleomorphs has recently been elegantly confirmed by Hansmann et al. (1986) using a monoclonal antibody against DNA and gold-labelled secondary antibody.

The many nuclear characteristics of the nucleomorph leave no doubt that it is a vestigial nucleus and prove that the chloroplast of cryptomonads has evolved from a symbiotic eukaryotic alga. What was this alga? The presence of phycobiliproteins in cryptomonads which are closely related to those in red algae indicates that the chloroplast evolved from a symbiotic red alga or an ancestor of present-day red algae. However, phycoerythyrin in cryptomonads is localized inside the thylakoids, not outside in phycobilisomes (Ludwig and Gibbs, 1989a; Rhiel et al., 1989). We have suggested a scenario for how this could have occurred (Ludwig and Gibbs, 1989a). In red algae, the genes for the phycobiliproteins and the anchor polypeptide are located on chloroplast DNA, whereas the genes for the linker polypeptides which hold the phycobilisomes together are located on nuclear DNA (Egelhoff and Grossman, 1983; Grossman et al., 1986). During the evolution of the cryptomonad chloroplast, a large part of the nucleomorph DNA has been lost or transferred to the nucleus (Ludwig & Gibbs, 1985), so it is not unlikely that the linker genes were lost. If this happened in a cell in which a mutation had occurred directing phycoerythrin to the thylakoid lumen, such a cell would survive in the absence of phycobilisomes. The question of which genes are located on nucleomorph DNA is currently of much interest. It is likely that the genes for the RNAs of the eukaryotic ribosomes in the periplastidal compartment are located in the nucleomorph (McKerracher and Gibbs, 1982). The gene for the small subunit of ribulose bisphosphate carboxylase would not be expected to be located on nucleomorph DNA, for in cryptomonads this protein is encoded by chloroplast DNA (Douglas and Durnford, 1989).

In 1930 Geitler discovered an unusual amoeboid alga which formed large plasmodia of amoeboid cells linked to each other by reticulopodia. The cells contained bright green chloroplasts but otherwise were quite unlike green algae. Geitler (1930) named this enigmatic protist, _Chlorarachnion reptans_. In 1984, Hibberd and Norris looked at _C. reptans_ with the electron

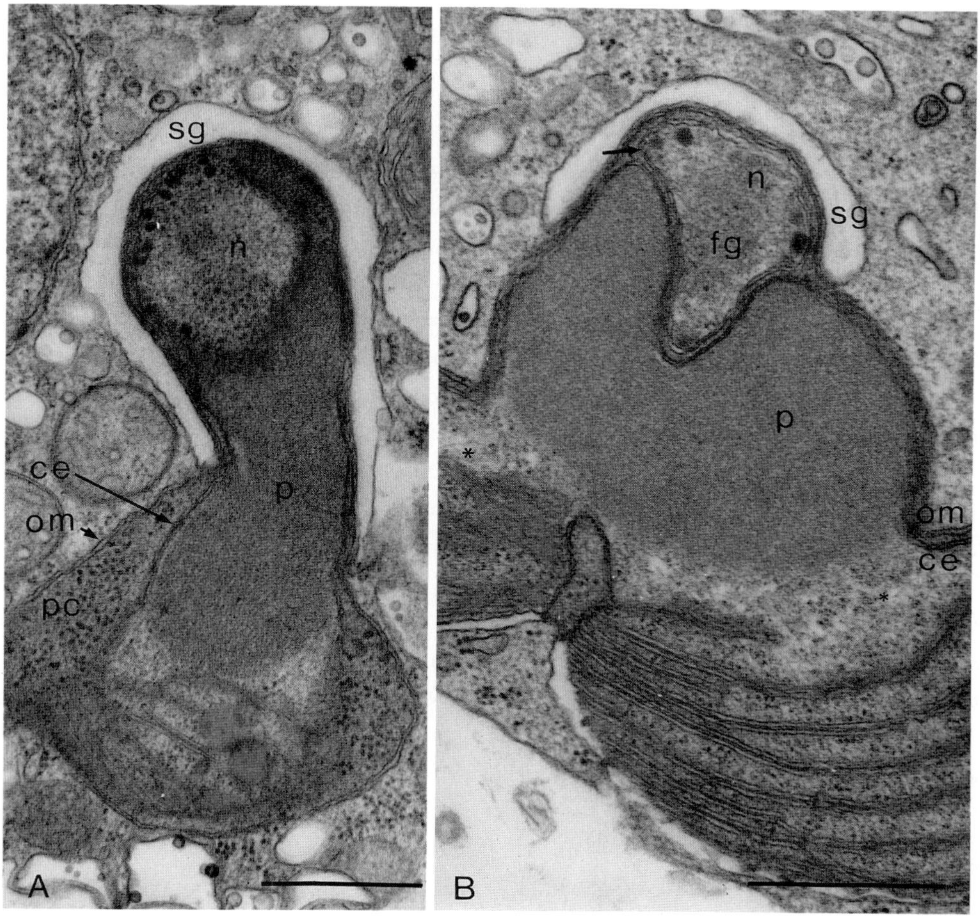

Fig. 3A,B. Cross-sections through the chloroplast of Chlorarachnion reptans. The chloroplasts are limited by four membranes, two of the chloroplast envelope (ce) and an outer membrane pair (om). Each chloroplast has a nucleomorph (n) associated with it located in a groove in the pyrenoid (p). A membrane-limited storage granule (sg) is present in the cytoplasm adjacent to the pyrenoid. A. Numerous eukaryotic-sized ribosomes are present in the periplastidal compartment (pc). B. The nucleomorph envelope is interrupted by pores (arrow). Fine fibrils of DNA can be seen in the nucleomorph matrix surrounding the fibrillogranular body (fg) as well as in the chloroplast stroma (asterisk). Scale bars = 0.5 μm

microscope and discovered that its chloroplasts were surrounded by four membranes and that each chloroplast had a nucleomorph-like body embedded in its pyrenoid. Figs. 3A and 3B are sections through the pyrenoid region of the chloroplasts of this unusual organism. Each chloroplast is surrounded by two pairs of membranes and the space between the outer pair and the chloroplast envelope contains abundant ribosomes, which are the same size as the eukaryotic ribosomes in the cell's cytoplasm (Fig. 3A). Note that unlike chloroplast ER, no ribosomes are present on the cytoplasmic surface of the outermost membrane (Fig. 3A,B). A nucleomorph, which is surprisingly similar morphologically to the nucleomorph of cryptomonads, is present in the periplastidal compartment lying in a groove in the pyrenoid. A membrane-limited carbohydrate storage granule caps the pyrenoid and nucleomorph exterior to the outer pair of membranes (Fig. 3A,B). The nucleomorph is surrounded by a double membrane which is interrupted by pores (arrow, Fig. 3B) and contains a fibrillogranular body (Fig. 3B) and rod-shaped dense globules. We have recently shown that the nucleomorph exhibits DNase-sensitive DAPI fluorescence (Ludwig & Gibbs, 1989b). Prominent DNA-like fibrils are seen in the nucleomorph matrix after Ryter and Kellenberger fixation (Ludwig and Gibbs, 1987). The nucleomorph divides in two in the same manner as the cryptomonad nucleomorph (Ludwig and Gibbs, 1989b). It is obvious that Chlorarachnion has acquired its chloroplast in the same manner as Cryptomonas by the reduction of a eukaryotic endosymbiont, only in this case the eukaryotic symbiont most likely was a green alga, since C. reptans contains chlorophylls a and b (Hibberd and Norris, 1984; C. Wilhelm, personal communication).

Recently, a green dinoflagellate has been discovered which contains chlorophylls a and b (Watanabe et al., 1987). The chloroplasts of this alga are limited by a double-membraned envelope and are separated from the dinoflagellate's cytoplasm by another pair of membranes, which, like those of Chlorarachnion, does not have ribosomes attached to the outermost membrane. The periplastidal compartment, between the outer pair of membranes and the chloroplasts, contains abundant ribosomes and several double-membrane-limited bodies. Watanabe et al. (1987) suggest that the double-membrane-limited bodies which have pores in their envelopes are nucleomorphs, although they have not yet been shown to contain DNA. Nonetheless, there is little doubt that this dinoflagellate has acquired its green chloroplasts from an endosymbiotic green alga.

Thus in three unrelated algae, cryptomonads, Chlorarachnion, and a green dinoflagellate, the chloroplast is surrounded by four membranes and in each case, the host organism has acquired its chloroplast by engulfing a eukaryotic alga and reducing it to a chloroplast. Each of these symbioses occurred independently of each other, but in each case, the cytoplasmic ribosomes and a probable remnant of the nucleus of the eukaryotic endosymbiont persist in the periplastidal compartment surrounding the chloroplast. In cryptomonads and Chlorarachnion, there is no doubt that the symbiont's nucleus persists, for the nucleomorph of each has been shown to contain DNA, a nucleolus-like body, and to divide in two amitotically prior to chloroplast and cell division. In fact, it is remarkable how the nucleus of the red algal symbiont in cryptomonads and the nucleus of the green algal symbiont in Chlorarachnion have evolved into such morphologically similar structures.

Because in these organisms, four membranes remain around the chloroplasts, I am convinced that in the seven other classes of algae which have chloroplasts surrounded by four membranes, the Prymnesiophyceae and the six heterokont classes (Raphidophyceae, Chrysophyceae, Bacillariophyceae, Phaeophyceae, Xanthophyceae and Eustigmatophyceae), the chloroplasts have evolved from endosymbiotic eukaryotic algae. Cavalier-Smith (1986) has argued that the chloroplast ER found in these groups and in cryptomonads is such a distinctive structure it could have evolved only once. He argues that this happened in the cryptomonads when they engulfed a chlorophyll c-containing eukaryotic alga and reduced it to a chloroplast. The heterokont algae and the Prymnesiophytes, he believes, evolved from the cryptomonads. I believe that this is unlikely. I do not believe that the cryptomonads with their characteristic periplast, unique trichocysts and distinctive rootlet system are related to the heterokont and Prymnesiophyte algae. The sequences of their cytoplasmic large subunit ribosomal RNA (Perasso et al., 1989) and 5S ribosomal RNA (Ludwig and Gibbs, 1985) indicate that they are not closely related to these algae. I do not believe it would be difficult for chloroplast ER to evolve more than once; we have seen that chloroplasts enclosed by four membranes have evolved independently at least three different times. For the outer pair of membranes to become chloroplast ER, the outer membrane would need only to acquire ribosomes by fusing with the nuclear envelope or rough ER of the host cell (Cavalier-Smith, 1986). Since Prymnesiophyte algae with their two equal flagella devoid of mastigonemes and their unique haptonema are very different from heterokont algae, although the chloroplasts of each group have many ultrastructural and biochemical similarities (Gibbs, 1970; Perasso et al., 1989), I believe that the Prymnesiophytes and heterokonts independently acquired their chloroplast by phagocytosing an eukaryotic alga, but in this case, the same eukaryotic alga became the symbiont. This alga had double-membrane-limited chloroplasts with chlorophylls _a_ and _c_, but no phycobiliproteins. However, this must remain a hypothesis, for such an alga does not exist today.

There remains only one class of algae to discuss, the dinoflagellates. Dinoflagellates are a large group of heterotrophic protists which have acquired chloroplasts in a variety of ways. Two dinoflagellates, _Peridinium balticum_ (Tomas and Cox, 1973) and _Glenodinium foliaceum_ (Jeffrey and Vesk, 1976), contain an entire Chrysophyte alga which occupies much of the cell. The symbiont lacks flagella, but has mitochondria, chloroplasts enclosed in chloroplast ER, abundant ribosomes, and a large eukaryotic nucleus, which appears normal, but divides amitotically (Tippit and Pickett-Heaps, 1976). A single membrane, the presumed plasmalemma of the eukaryotic endosymbiont, separates the symbiont's cytoplasm from the cytoplasm of the host dinoflagellate. Recently, Wilcox and Wedemayer (1984) have discovered that _Gymnodinium acidotum_ contains an endosymbiotic cryptomonad cell. However, unlike _Peridinium balticum_ and _Glenodinium foliaceum_, _Gymnodinium acidotum_ has not been grown in culture, and Larsen (1988) has suggested that it is only feeding on cryptomonads. Larsen (1988) has described a dinoflagellate, _Amphidinium poecilochroum_, which feeds on cryptomonads by penetrating its prey with a feeding tube and sucking the cell's contents into a phagocytic vacuole, a process named myzocytosis by Schnepf and Deichgräber (1984). _A. poecilochroum_ contains cryptomonad cells which lack trichocysts and periplast plates and are enclosed by a phagocytic vacuole membrane. These endosymbionts are very similar to those found in _Gymnodinium acidotum_, but the cryptomonad endosymbionts in _A. poecilochroum_ are ultimately digested (Larsen, 1988). However, several other dinoflagellates appear to contain cryptomonad chloroplasts as permanent symbionts. In _Amphidinium wigrense_, the cryptomonad chloroplasts are surrounded by three membranes (Wilcox and Wedermayer, 1985). The third membrane may be the former plasmalemma of a cryptomonad endosymbiont or, alternatively, the phagocytic vacuole membrane if the cryptomonad chloroplasts were acquired by myzocytosis. _Dinophysis acuta_ and _D. acuminata_ contain cryptomonad chloroplasts surrounded by only two membranes (Schnepf and Elbrächter, 1988). These chloroplasts are presumed to have been taken up by myzocytosis and the phagocytic vacuole membrane lost. I have already discussed the green dinoflagellate which contains green algal chloroplasts surrounded by four membranes. In this case, the green algal symbiont was taken up by phagocytosis, and both the plasmalemma of the green alga and the phagocytic vacuole membrane of the dinoflagellate persist.

These dinoflagellates are all unusual. Most dinoflagellates possess chloroplasts which are surrounded by three membranes and contain chlorophylls a and c. How these chloroplasts have evolved remains a mystery. They may have evolved directly from a symbiotic prokaryote, perhaps from Raven's (1970) hypothetical chlorophyll a- and c-containing prokaryote. However, the presence of a third membrane surrounding the chloroplasts suggests that these chloroplasts also may have evolved from symbiotic eukaryotic algae. The third membrane may be derived from the plasmalemma of the symbiont (Gibbs, 1978) or from the phagocytic vacuole membrane of the host cell if the dinoflagellate ingested its prey by myzocytosis (Schnepf and Deichgräber, 1984; Schnepf and Elbrächter, 1988). But which eukaryotic alga was endocytosed is not known. We have seen in Fig. 1 that the sequences of cytoplasmic ribosomal RNAs are giving us a better understanding of how the different groups of eukaryotes are related to each other. We now need sequence data for the ribosomal RNAs of the chloroplasts in each of the classes of algae in order to determine the evolutionary history of algal chloroplasts more accurately. Since two symbioses have been involved in the evolution of most algal chloroplasts, and in some cases three symbioses (the cryptomonad chloroplasts in dinoflagellates), constructing chloroplast evolutionary trees will be a challenging task.

Acknowledgements. This work was supported by a grant from the Natural Sciences and Engineering Research Council of Canada. I wish to thank M. Ludwig for the electron micrographs and R. Lamarche for the dendrogram.

References

Bhattacharya D, Druehl LD (1988) Phylogenetic comparison of the small-subunit ribosomal DNA sequence of *Costaria costata* (Phaeophyta) with those of other algae, vascular plants and oomycetes. J Phycol 24: 539 - 543

Bonen L, Doolittle WF (1975) Proc Natl Acad Sci USA 72: 2310 - 2314

Cavalier-Smith T (1982) The origin of plastids. Biol J Linn Soc 17: 289 - 306

Cavalier-Smith T (1986) The Kingdom Chromista: Origin and Systematics. In: Round FE, Chapman DJ (eds) Progress in Phycological Research 4, Biopress, Bristol, pp 309 - 347

Douglas SE, Durnford DG (1989) The small subunit of ribulose-1,5-bisphosphate carboxylase is plastid-encoded in the chlorophyll c-containing *Cryptomonas sp.* Plant Mol Biol 13: 13 - 20

Egelhoff T, Grossman A (1983) Cytoplasmic and chloroplast synthesis of phycobilisome polypeptides. Proc Natl Acad Sci USA 80: 3339 - 3343

Geitler L (1930) Ein grünes Filarplasmodium und andere neue Protisten. Arch Protistenk 69: 615 - 636

Gibbs SP (1970) The comparative ultrastructure of the algal chloroplast. Ann NY Acad Sci 175: 454 - 473

Gibbs SP (1978) The chloroplast of *Euglena* may have evolved from symbiotic green alga. Can J Bot 56: 2883 - 2889

Gibbs SP (1981) The chloroplast endoplasmic reticulum, structure, function, and evolutionary significance. Int Rev Cytol 72: 49 - 99

Gillott MA, Gibbs SP (1980): The cryptomonad nucleomorph: its ultrastructure and evolutionary significance. J Phycol 16: 558 - 568

Gray MW, Doolittle WF (1982) Has the endosymbiont hypothesis been proven? Microbiol Rev 46: 1 - 42

Greenwood AD, Griffiths HB, Santore UJ (1977): Chloroplasts and cell compartments in Cryptophycae. Brit Phycol J 12: 119

Grossman AR, Lemaux PG. Conley PB (1986): Regulated synthesis of phycobilisome components. Photochem Photobiol 44: 827 - 837

Gunderson JH, Elwood H, Ingold A, Kindle K, Sogin ML (1987) Phylogenetic relationships between chlorophytes, chrysophytes, and oomycetes. Proc Natl Acad Sci USA 84: 5823 - 5827

Hansmann P (1986) Ultrastructural Localization of RNA in Cryptomonads. Protoplasma 146: 81 - 88

Hansmann P, Falk H, Scheer U, Sitte P (1986) Ultrastructural localization of DNA in two *Cryptomonas* species by use of monoclonal antibodies. Eur J Cell Biol 42: 152 - 160

Hansmann P, Falk H, Sitte P (1985) DNA in the nucleomorph of *Cryptomonas* demonstrated by DAPI fluorescence. Z Naturforsch 40c: 933 - 935

Hansmann P, Maerz M, Sitte P (1987) Investigations on genomes and nucleic acids in cryptomonads. Endocyt Cell Res 4: 289 - 295

Hibberd DJ, Norris RE (1984) Cytology and ultrastructure of *Chlorarachnion* reptans (Chlorarachniophyta divisio nova, Chlorarachniophyceae classis nova). J Phycol 20: 310 - 330

Jeffrey SW, Vesk M (1976) Further evidence for a membrane-bound endosymbiont within the dinoflagellate *Peridinium foliaceum* . J Phycol 12: 450-455

Larsen J (1988) An ultrastructural study of *Amphidinium poecilochroum* (Dinophyceae) a phagotrophic dinoflagellate feeding on small species of cryptophytes. Phycologia 27: 366 - 377

Loeblich AR III (1976) Dinoflagellate Evolution: Speculation and Evidence. J Protozool 23: 13 - 28

Ludwig M, Gibbs SP (1985) DNA is present in the Nucleomorph of Cryptomonads: Further Evidence that the Chloroplast Evolved from a Eukaryotic Endosymbiont. Protoplasma 127: 9 - 20

Ludwig M, Gibbs SP (1987): Are the nucleomorphs of cryptomonads and *Chlorarachnion* the vestigial nuclei of eukaryotik endosymbionts. Ann NY Acad Sci 503: 198 - 211

Ludwig M, Gibbs SP (1989a) Localization of Phycoerythrin at the Lumenal Surface of the Thylakoid Membrane in *Rhodomonas lens*. J Cell Biol 108: 875 - 884

Ludwig M, Gibbs SP (1989b) Evidence that the nucleomorphs of *Chlorarachnion reptans* (Chlorarachniophyceae) are vestigial nuclei: Morphology, division and DNA-DAPI fluorescence. J Phycol 25: 385 - 394

McKerracher L, Gibbs SP (1982) Cell and nucleomorph division in the alga *Cryptomonas*. Can J Bot 60: 2440 - 2452

Morden CW, Golden SS (1989) psbA genes indicate common ancestry of the prochlorophytes and chloroplasts. Nature (Lond) 337: 382 - 384

Morrall S, Greenwood AD (1982) Ultrastructure of nucleomorph division in species of Cryptophyceae and its evolutionary implications. J Cell Sci 54: 311 - 328

Perasso R, Baroin A, Qu LH, Bachellerie JP, Adoutte A (1989) Origin of the algae. Nature (Lond) 339: 142 - 144

Raven P (1970) A Multiple Origin of Plastids and Mitochondria. Science 169: 641 - 646

Rhiel E, Kunz J, Wehrmeyer W (1989) Immunocytochemical Localization of Phycoerythrin-545 and of a Chlorophyll a/c Light Harvesting Complex in *Cryptomonas maculata* (Cryptophyceae). Botanica Acta 102: 46 - 53

Schnepf E, Deichgräber G (1984) "Myzocytosis", a Kind of Endocytosis with Implications to compartmentation in Endosymbiosis. Observations in *Paulsenella* (Dinophyta). Naturwiss 71: 218 - 219

Schnepf E, Elbrächter M (1988) Cryptophycean-Like Double Membrane-Bound Chloroplast in the Dinoflagellate, *Dinophysis* Ehrenb.: Evolutionary, Phylogenetic and Toxicological Implications. Botanica Acta 101: 196 - 203

Sogin ML, Elwood HJ, Gunderson JH (1986) Evolutionary diversity of eukaryotic small-subunit rRNA genes. Proc Natl Acad Sci USA 83: 1383 - 1387

Sogin ML, Gunderson JH, Elwood HJ, Alonso RA, Peattie DA (1989) Phylogenetic Meaning of the Kingdom Concept: An Unusual Ribosomal RNA from *Giardia lamblia*. Science 243: 75 - 77

Tippit DH, Pickett-Heaps JD (1976) Apparent amitosis in the binucleate dinoflagellate *Peridinium balticum*. J Cell Sci 21: 273 - 389

Thomas RN, Cox ER (1973) Observations on the symbiosis of *Peridinium balticum* and its intracellular alga. I. Ultrastructure. J Phycol 9: 304 - 323

Turner S, Burger-Wiersma T, Giovannoni SJ, Mur LR, Pace NR (1989) The relationship of the prochlorophyte *Prochlorothrix hollandica* to green chloro- plasts. Nature (Lond) 337: 380 - 382

Van den Eynde H, De Baere R, De Wachter R (1988) Sequence and secondary structure of *Porphyra umbilicalis* 5S rRNA. Relevance for the evolutionary origin of red algae. Nucleic Acids Res 16: 10919

Watanabe MM, Takeda Y, Sasa T, Inouye I, Suda S, Sawaguchi T, Chihara M (1987) A green dinoflagellate with chlorophylls a and b: Morphology, fine structure of the chloroplast and chlorophyll composition. J Phycol 23: 382 - 389

Wilcox LW, Wedemayer GJ (1984) *Gymnodinium acidotum* Nygaard (Pyrrophyta), a dinoflagellate with an endosymbiotic cryptomonad. J Phycol 20: 236 - 242

Wilcox LW, Wedemayer GJ (1985) Dinoflagellate with Blue-Green Chloroplasts Derived from an Endosymbiotic Eukaryote. Science 227: 192 - 194

Woese CR (1987) Bacterial evolution. Microbiol Rev 51: 221 - 271

Phycobilisomes: Structure and function

Werner Wehrmeyer
Fachbereich Biologie
Universität Marburg
Karl-von-Frisch-Straße
D 3550 Marburg
Federal Republic of Germany

Light harvesting pigments in the photosynthetic apparatus of plants function in light absorption and transfer of excited state energy to the reaction centres of the photosystems. In the evolution of antenna pigment systems in photoautotrophic plants the development of phycobiliproteins has played a decisive role. 1. The phycobiliproteins are present in the prokaryotic cyanobacteria which are virtually the first oxygen producing photosynthetic organisms, arising about 3 billion years ago (33). 2. There is increasing evidence that many recent species of cyanobacteria are capable not only of oxygenic, but also of (facultative) anoxygenic photosynthesis under appropriate conditions (26). This ability is reminiscent of the corresponding step from anaerobic to microaerobic life in the evolution of photosynthetic organisms in the Precambrium era. 3. Structurally and biochemically the chloroplasts of eukaryotic red algae show a high similarity with the photosynthetic apparatus of cyanobacteria and it is suspected have evolved from a cyanobacterium after an endosymbiotic event. Phycobiliproteins have thus successfully survived the evolutionary transition from pro- to eukaryotic cells (4o). 4. Measurements of the photosynthetic action spectra of cyanobacteria and red algae show that most of the light responsible for oxygen production is captured by phycobiliproteins in an absorption range between 45o and 65o nm, whereas wavelengths absorbed by chlorophyll and carotenoids make little or no contribution. The functional importance of phycobiliproteins is therefore well established (2).

Phycobiliproteins are organized in cyanobacteria and red algae as supramolecular, multiprotein structures called phycobilisomes (= pbsomes). Pbsomes have a molecular mass in the range of 5 to 2ox10^{6} da and are therefore easily visible in the electron microscope. They are associated at the outer surface of the thylakoid membranes and - like the chlorosomes in the green photosynthetic bacteria - represent peripheral antenna systems. This location and the hydrophilic character of the biliproteins has greatly facilitated their isolation in pure fractions with detergents using high molar phosphate buffer for further stabilization. Thus combined structural and functional studies were possible.

Morphology and architecture of phycobilisomes

Different types of pbsomes have been described (39). In cyanobacteria the hemidiskoidal pbsomes with a small profile (1onm) are predominant; in red algae the same type can also be present, but hemiellipsoidal pbsomes about 2o nm thick become predominant particularly in higher taxonomic orders. If evolution is looked upon as a process of functional optimizing and increase of complexity, the bigger pbsomes of red algae may be the more effctive in light absorption, energy transfer or regulation of energy distribution. This is at present not completely understood. Until recently, only the architecture of hemidiskoidal pbsomes had been analyzed in detail. The model of hemidiskoidal pbsomes of Rhodella violacea proposed in 1977 (2o) is generally accepted. With slight modifications it also fits for cyanobacterial pbsomes (3, 23) (Fig. 1).
A triangular core consisting of three cylindrical elements (11 nm Ø) is surrounded on two sides by six rods each made up of stacks of disk-shaped units (1o-11 nm Ø, 5-6 nm thick). These rods show strong repeats at 6 nm and weaker ones at 3 nm intervals perpendicular to their axes, indicating hexameric and trimeric biliprotein aggregates, respectively.
In cooperation with Dr. K. Ohki recently we discovered a new type of cyanobacterial pbsome taking an intermediate position between hemidiskoidal and hemiellipsoidal types (42). The model of the new pbsome in face view (Fig. 2) shows its extraordinary size: 65-8o nm wide at the base, 35-4o nm in height, distal thickness 2o nm, proximal at the attachment point with

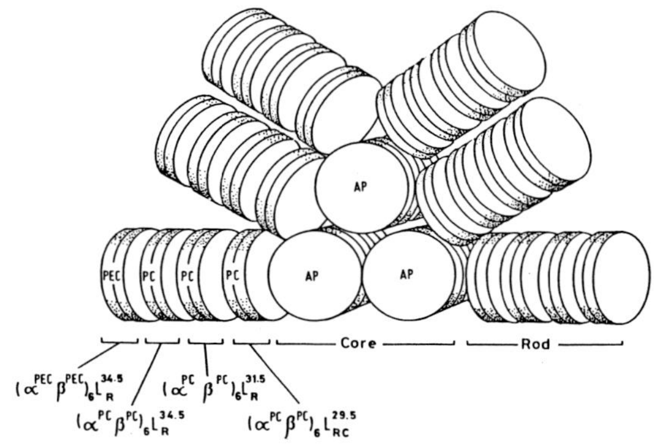

Fig. 1. Phycobilisome model of the cyanobacterium Mastigocladus laminosus (23, 45). Abbreviations: Phycoerythrocyanin PEC, Phycocyanin PC, Allophycocyanin AP, Linker polypeptide L; the subscripts R, C, RC or CM denote the positions of the linkers in the rod or core substructure or at the rod-core or core-membrane interface. The superscripts denote their apparent molecular masses in kDa.

the thylakoid membrane 1o nm. The triangular centre is similar to that in Fig. 1, the periphery, however, has bundles of 5 laterally associated rods attached to two sides of the core. The uppermost cylinders form an obtuse angle and the lowermost two cylinders are inclined to the supporting thylakoid membrane. On the basis of a 7:1:1 molar ratio of PE : PC : AP a molecular mass of $14.6x10^6$ Da could be calculated.

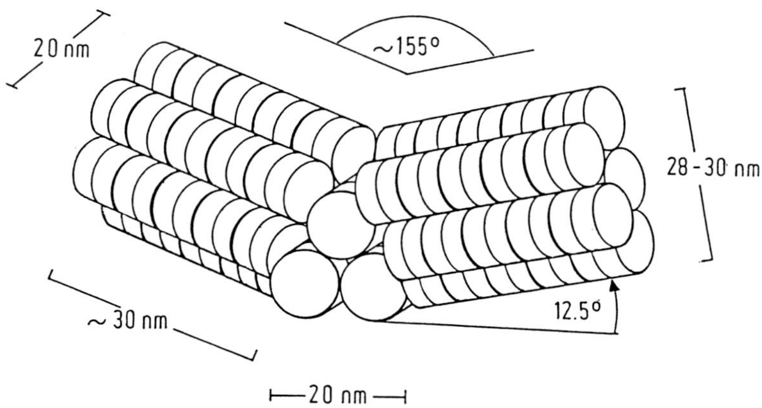

Fig. 2. Phycobilisome model of the cyanobacterium Phormidium persicinum (42).

The sedimentation constant was determined to be 98 S. It can be demonstrated (Table I) that in both cyanobacteria and red algae there is an increasing concentration of bilin chromophores in the pbsome periphery. This parallels the morphological progression from hemidiskoidal to hemiellipsoidal pbsomes in the red algae and is similarly expressed in the step from the hemidiskoidal to the newly described Phormidium type of pbsomes in the cyanobacteria.

SOURCE	MOLAR RATIO OF BILIPROTEINS	RATIO OF CHROMOPHORES	CALCULATED MOLECULAR MASS OF PHYCOBILISOME (DALTON)	SEDIMENTATION CONSTANT	PHYCOBILISOME DIMENSION (NM)	REF.
CYANOBACTERIA						
SYNECHOCYSTIS 6701 (GREEN LIGHT)	C-PE 3, C-PC 1, AP 1	PEB 540 PCB 108 PCB 72	$7\text{-}8 \cdot 10^6$ (GREEN) $4.5 \cdot 10^6$ (RED)	47 N.D.	35 x 28 x 10	(4) (44)
PHORMIDIUM PERSICINUM	C-PE 7.3, C-PC 1, AP 1	PEB 1320 PCB 108 PCB 72	$14.6 \cdot 10^6$	98	70 x 35 x 21	(25) (42)
RHODOPHYTA						
RHODELLA VIOLACEA	B-PE 2, C-PC 1, AP 1	PEB + PUB 408 PCB 108 PCB 72	$5.6\text{-}6 \cdot 10^6$		36 x 29 x 8	(16) (42)
PORPHYRIDIUM CRUENTUM	B-PE 15, R-PC 2, AP 1	PEB + PUB 2040 PEB + PCB 144 PCB 48	$19\text{-}22 \cdot 10^6$	126	40 x 27 x 19 50 x 30 x 22	(7) (25) (39)

Table I. Increase in bilin chromophores of the pbsome periphery in parallel to the progression of the pbsome type from hemidiskoidal to hemiellipsoidal, which is obtained in red algae, but not in cyanobacteria. Phycoerythrobilin PEB, Phycourobilin PUB, Phycocyanobilin PCB.

The increase of absorbing capacity in the periphery can be traced back, in part, to the numerical increase of phycoerythrin containing aggregates, basically, to the high number of bilin chromophores bound to α and ß subunits of phycoerythrin hexamers, i.e. 2 or 3, respectively (15).

Biochemical composition of pbsomes and dynamic variation

The major components of pbsomes are the bilin-containing proteins PE, PC and AP. In some cyanobacteria PE can be replaced by PEC or both pigments can be lacking; PC and AP, however, are constitutively present in all cyanobacteria and red algae studied (4, 7). SDS gel electrophoreses of isolated pbsomes reveal the presence of α and ß subunits of the biliproteins in the 14 to 22 kDa range making up 85% of the total protein; in addition about 15% linker polypeptides are present (9, 36).

The latter can be classified into 4 groups related to their function and molecular weights: I. 7o-12o, II. 25-7o, III. 15-22, IV. 8-12 kDa. Linker polypeptides of group I. are responsible for the attachment of the pbsomes to the thylakoid membrane and the spatial organization of the tricylindrical core. In the AN 112 mutant of Synechococcus 63o1 the pbsomes contain an unusual bicylindrical core (19), probably as a result of the reduced linker L_{CM}^{75}. Linkers of group I. have been shown to contain a phycocyanobilin chromophore which enables them to carry out their role as terminal acceptors of excitation energy in the pbsomes (18, 27). Group II linker polypeptides function in the sequential arrangement of biliprotein aggregates in the peripheral rods and determine the specificity of their location (Fig. 1). In particular, they organize face to side coupling of proximal constitutive phycocyanin aggregates to the allophycocyanin trimers of the core. In this process of initiating rod assembly they are perhaps supported by linkers of group III. As shown for B-phycoerythrin (22) linkers of group II. are probably partially buried within the central channel of hexameric biliprotein aggregates. This enables linker polypeptides to specifically influence the phycocyanobilin chromophore at ß C-84, located in this area, and thus change spectral and energy transfer characteristics. As linkers associated with B-phycoerythrin carry chromophoric groups themselves they can contribute to energy transfer directly. The modulation of the spectroscopic properties of biliprotein aggregates is also a characteristic feature of the L_C^{1o} linker of group IV in allophycocyanin aggregates (28), however, an additional function as rod terminating protein has been suggested (6). Pbsomes are adaptable structures influenced quantitatively and qualitatively by various environmental factors, especially by light intensity, light quality and nutrient conditions. The total biliprotein content of the cells is inversely related to the total growth irradiance levels: Species cultured under high light have the lowest, under low light the greatest amount of biliprotein pigment per cell and thus the highest number of pbsomes per μm^2. Well known is the complementary chromatic adaptation of some cyanobacteria; this implies that under red light, phycoerythrin is replaced by phycocyanin whereas under green light, phycoerythrin is favoured instead of phycocyanin (38). This bidirectional complementary chromatic adaptation has so far only been described for hemidiskoidal pbsomes. This type is preserved when the spectral characteristics and the effective absorption cross section change as a consequence of the altered biliprotein composition. The sedimentation constant may be similar or slightly reduced after red light (4, 24). Studying a chromatically adapting Phormidium sp. we recently found that chromatic adaptation can also lead to completely different pbsome types within one species. Under green light conditions a molar ratio of PE : PC : AP of 7.3 : 1 : 1 was determined. The pbsomes had a molecular mass of 14.6×10^6 Da, a sedimentation constant of 98 S, and showed 1o peripheral rods as had been demonstrated in Ph. persicinum (Fig. 2). Under red light, the molar ratio of the constitutive biliproteins was 1.2 : 3.3 : 1, the calculated molecular mass was 8.5×10^6 Da, the sedimentation constant 66 S and the pbsomes were of the hemidiskoidal type with 6 rods. SDS gel electrophoreses of pbsomes isolated from Phormidium sp. after culture under green and red light confirm the prevailing changes in the biliprotein composition but, in addition, substantial changes in the linker polypeptides of groups II, IV and, in particular, group I occur. These unexpected changes of core constituents required further investigations, especially, improved isolation procedures of core fractions. Dissociated pbsomes were separated by gel filtration and the fractions subsequently purified by either gel electrophoresis or sucrose density gradient centrifugation. Table II. summarizes the results. Among the quickly dissociating allophycocyanin-containing constituents of the core in the range of 1oo - 12o kDa, the AP 68o complex is the most remarkable. Only this complex contains α^{AP-B} and is responsible for the long wavelength fluorescence emission of pbsomes at room temperature near 68o nm (77 K emission at 685 nm) (28). In the pbsome models proposed so far, two copies each of $(\alpha^{AP}\beta^{AP})_3$ and $(\alpha^{AP}\beta^{AP})_3 \cdot L_C^{1o}$ are presumed to form the apical cylinder; in addition, two of each of these complexes make up the periphery

	ABSORPTION MAX. (NM)	FLUORESCENCE EMISSION (NM)	SUBUNIT COMPOSITION	COMPLEX ORGANIZATION	$M_R 10^{-3}$
AP 660	651 620s, 594s	20°C 662 77 K 664	α^{AP}, β^{AP}	$(\alpha^{AP}\beta^{AP})_3$	105.0
PC 646	620	20°C 646 77 K 646	α^{PC}, β^{PC}	$(\alpha^{PC}\beta^{PC})_3$	118.5
PC 642	620	20°C 640 77 K 642	$\alpha^{PC}, \beta^{PC}, L_R^{11}$	$(\alpha^{PC}\beta^{PC})_3 L_R^{11}$	129.5
AP 664	653 619s, 594s	20°C 664 77 K 664	$\alpha^{AP}, \beta^{AP}, L_C^{10}$ $\alpha^{*AP}, \beta^{*AP}$	$(\alpha^{*AP}\alpha_2^{AP}\beta_2^{AP}\beta^{*AP}) L_C^{10}$	115.0
AP 680	654 678s, 619s, 594s	20°C 680 77 K 686	$\alpha^{AP}, \beta^{AP}, L_C^{10}$ $\alpha^{AP\cdot B}, \alpha^{*AP}, \beta^{*AP}$	$(\alpha^{*AP}\alpha^{AP}\alpha^{AP\cdot B}\beta_2^{AP}\beta^{*AP}) L_C^{10}$	115.2
AP_C	654,620s	20°C:670,680 77 K:685,670s	$\alpha^{AP}, \beta^{AP}, \alpha^{APB}, L_C^{10}$ $\alpha^{*AP}, \beta^{*AP}, \beta^{19.5}, L_{CM}^{105}$	$(\alpha^{*AP}\alpha_2^{AP}\beta_2^{AP}\beta^{*AP})_4 . \beta_2^{19.5} (L_{CM}^{105})_2$ 2 x AP 680 2 x AP 664	$1 \cdot 10^3$
AP_{CM}	654,620s	20°C:671 77 K:673	α^{AP}, β^{AP} $\alpha^{*AP}, \beta^{*AP}, \beta^{19.5}, L_{CM}^{105}$	$(\alpha^{*AP}\alpha_2^{AP}\beta_2^{AP}\beta^{*AP})_6 . \beta^{19.5} (L_{CM}^{105})_2$	850

Table II. Composition and characteristic features of biliprotein complexes of the 1oo-12o kDa fraction (above) and of the high molecular core fraction ($\geq$ 85o kDa, below) of dissociated pbsomes of M. laminosus separated by gel filtration and subsequent gel electrophoresis or sucrose density gradient centrifugation (28, 29).

of basal cylinders of tricylindrical cores. Two long-wavelength emitting complexes $(\alpha^{AP-B}\alpha_2^{AP}\beta_3^{AP}) \cdot L_C^{1o}$ and $(\alpha^{AP}\beta^{AP})_2 \cdot \beta^{18.5} \cdot L_{CM}^{99}$, on the other hand, are suggested to constitute the central part of the two basal cylinders of the core (9,1o). The easy and quick release of AP 68o from the core together with the trimeric complexes AP 66o and AP 664 from partially dissociated pbsomes of M. laminosus contradicts a central location. Thus a peripheral position of AP 68o is presumed (Fig. 3 a, b). This is supported by the detection and purification of a high molecular native fraction equivalent to about 5o% allophycocyanin of the centre; it contains all high molecular linker polypeptide L_{CM}^{1o5} and is completely devoid of AP 68o, the linker L_C^{1o} and any traces of phycocyanin. The criteria of purity were: 1. the complete lack of contaminating protein bands in SDS gel electrophoreses and 2. the quality of spectral curves enhanced by secondary derivative spectroscopy. Under these purity conditions, the fluorescence emission maximum at room temperature lies at 671 nm, the 77 K fluorescence emission at 673 nm. These characteristics have never before been shown for a native core complex. Since this complex presents the organization centre of the pbsome core and the connecting link to the supporting thylakoid membrane, it was designated AP_{CM} (29). If trimeric allophycocyanin complexes AP 664 and AP 68o are associated at the faces of AP_{CM} to form a high molecular complex according to Fig.3, a dual fluorescence emission with maxima at 67o and at 68o nm can be observed at room temperature. These emission maxima are indicative of two functionally independent, although spatially associated, terminal emitters. The emission at 68o nm (25 ^{o}C) and at 685 nm (77 K) can well be assigned to allophycocyanin-B within the AP 68o complex and the 67o nm emission can be correlated with the L_{CM}^{1o5} linker polypeptide of the AP_{CM} (29).

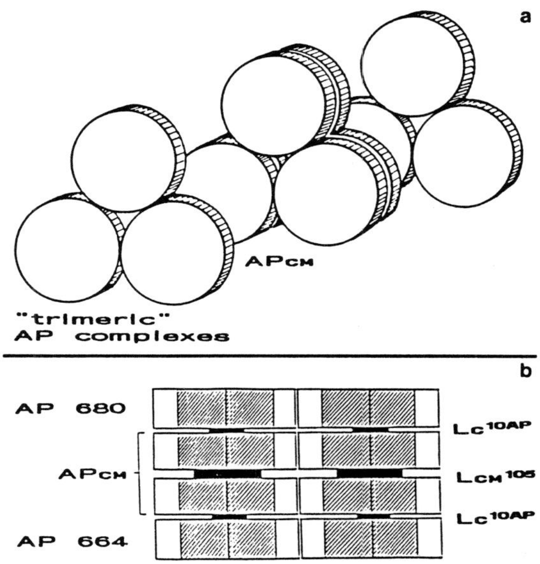

Fig. 3 a, b. Spatial arrangement of the constituents of the pbsome core of M. laminosus; a) Face view of the core with the peripheral trimeric AP complexes extended for better visualization of the central AP_{CM}; b) Edge view of the basal cylinders with the presumed distribution of the linker polypeptides (black) (29). Abbreviations as in Fig. 1

Energy transfer

In spite of the fact that proteins and especially aromatic amino acids in proteins can contribute to and enhance energy transfer the decisive role is performed by chromophores. Active in energy transfer in biliproteins are the bilin chromophores phycocyanobilin, phycoerythrobilin, phycourobilin and phycobiliviolin (1o). These are convalently bound by thioether bridges of one or two cysteinyl residues of the polypeptide chain with ring A or ring A and D of the bilin, respectively (17). The cysteinyl residue at C 84 in ß subunits of many different biliproteins constantly carries phycocyanobilin as the chromophore. This indicates a specific functional role in energy transfer. Other cysteinyl residues α C 84 and ß C 155 can carry different, singly bound chromophores; the residues ß C 5o and C 61 offer two linkages for doubly bound chromophores which also can be different. According to ideas of Förster on energy transfer the total transfer rate depends mainly on the relative distances between the chromophores of donor and acceptor molecules, their orientations, and the spectral overlap of the emission of the donor with the absorption spectrum of the acceptor (5). X-ray crystallographic analysis of trimeric and hexameric phycocyanin crystals yielded valuable data on relative distances and orientation factors (32). When viewed down the threefold axis phycocyanin trimers reveal the phycocyanobilin chromophores within the apoprotein conformation: ß 155 is located at the periphery, α 84 in the monomer/monomer contact region, whereas ß 84 stretches into the central channel inclined to the trimer plane at 30° (3o). The distances of chromophores (ß 155 to ß 84 or α 84 to ß 84) were determined to be 38.8 and 2o.8 Å, respectively. When hexamers are formed by stacking of two trimers, which predominantly takes place by interaction of polar and charged residues of α chains, the chromophores of the different trimers are separated by less than 37 Å (31). In contrast to the tight intrahexameric association of the trimers the interaction of hexamers in the rods forming higher aggregates is restricted to very few residues of the ß chain lying on the top. The interhexameric energy transfer takes place preferentially via the central ß 84 chromophores with a distance to the next ß 84 of 26 Å.

Numerous picosecond time-resolved studies on energy transfer dynamics within isolated pbsomes and their defined substructures have been performed during the last decade. Different detection techniques have been used (a) measurement of ground state recovery kinetics after picosecond pulse excitation and (b) measurement of fluorescence decay by the single photon timing (SPT) technique (12). The fastest processes observed by SPT in isolated pbsomes of Synechocccus and Porphyridium cruentum and allophycocyanin complexes of M. laminosus were in the order of 1o ps and may tentatively be assigned to intramolecular s-f transfer between neighbouring chromophores within phycocyanin or phycoerythrin hexamers or allophycocyanin trimers (1, 8, 35, 41, 43). This is supported by similar fast decays of induced absorption anisotropy measured in aggregates of Synechococcus and Porphyridium cruentum (8, 43). Measurements of excitation energy transfer with fluorescence decay times in the range of 45 to 12o ps could be attributed to the overall rod-core transfer increasing with rod length. With 4o-5o ps it is shorter in pbsomes of the Synechococcus 63o1 AN 112 mutant with only one PC hexamer per rod and is extended to 9o ps in the original Synechococcus 63o1 strain with 3 PC hexamers per rod (8). Fluorescence decays of about 6o ps (34 ps) and 4o ps (25 ps) were measured after selective excitation of B-phycoerythrin and phycocyanin in hemiellipsoidal pbsomes of P. cruentum and hemidiskoidal types of Rh. violacea (values in brackets) and are interpreted as energy-transfer times from phycoerythrin to phycocyanin and phycocyanin to allophycocyanin, respectively (12, 43). Two different models have been proposed for the energy transfer along the rods:(a) The energy transfer may be unidirectional along a linear array where the disk-to-disk transfer represents the rate limting step (11); (b) the energy may be equilibrated rapidly within the rods by fast interdisk transfer and the rate limiting step is at the face-to-side transition from rod to core complexes (13, 36).

Energy transfer within the pbsome core is less well understood (14). Allophycocyanin differs from the other biliproteins in many aspects: 1. Almost no residues which determine the trimer -trimer association are conserved; therefore, allophycocyanin does not form hexameric aggregates. 2. Threefold symmetry along the z axis, a characteristic feature of PE and PC aggregates (21, 22, 3o, 34), is lost, at least in AP 68o with one asymmetrically located α^{AP-B} subunit. Microheterogeneity within α and ß subunits of the AP 664 and AP 68o complexes may be indicative of internal inhomogeneity or loss of symmetry (29). 3. Polarity with an orientation to the thylakoid membrane specifically is obtained by the spatial incorporation of the L_{CM}^{1o5} linker polypeptide being exclusively located in the AP_{CM} core complex (29). Contrary to the energy fluxes within the peripheral rods following the longitudinal 3-fold axes the main energy flux within the core has to transverse the allophycocyanin aggregates at right angle to the 3-fold core axis. In Fig.4 this main energy flow coming from constitutive phycocyanin aggregates (PC) and indicated by double-lined arrows is funneled to the reaction centres of photosystem II via the central core complexes of the pbsome AP_{CM} and the core antenna complexes CP 43 and CP 47 of the reaction centre. An energy distribution mechanism horizontally orientated between pbsomes aligned within one row may take place via the peripheral core complexes AP 664 and AP 68o (one-lined arrow). Moreover, it cannot be excluded that the 68o nm emission of AP 68o is directly accepted by core antenna complexes of photosystem I.

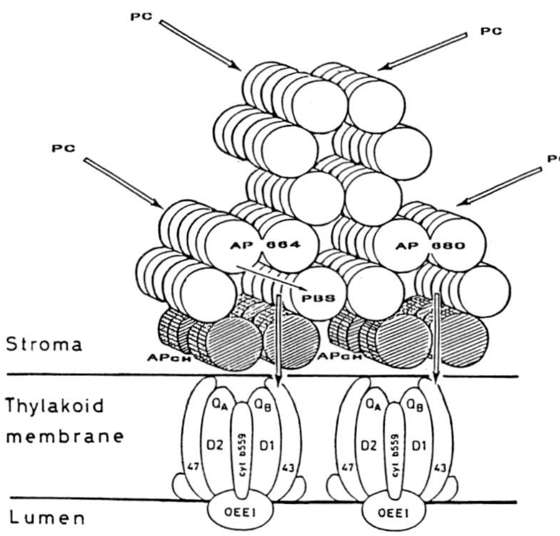

Fig. 4. Model of the cooperation of one hemidiskoidal phycobilisome (only the core is depicted) and two photosystem II complexes with the energy distribution pathways indicated by arrows. Energy flow from the peripheral rods (PC) polarly funneled via the two centrally located AP_{CM} complexes to the membrane-bound core antenna complexes CP_{43} and CP_{47} and the reaction centre of photosystem II $(D_1/D_2)^{43}$ (double-lined arrows). A horizontal energy distribution from AP 664 to neighbouring pbsomes in front or behind the plane may be an additional possibility (one-lined arrow); the emission from AP 68o directed to photosystem I as a bypass is not yet established and not shown (29).

The experimental work of our group incorporated in this report was supported by the Deutsche Forschungsgemeinschaft (Sonderforschungsbereich 3o5) and the Fonds der Chemischen Industrie.

References

1. Bittersmann E, Reuter W, Wehrmeyer W, Holzwarth AR (1988) Picosecond energy transfer kinetics in allophycocyanin aggregates from Mastigocladus laminosus. In: Scheer H, Schneider S (eds) Photosynthetic Light-Harvesting Systems, de Gruyter, Berlin New York, p 451-455
2. Blinks LR (1964) Accessory pigments and photosynthesis. In: Giese AC (ed) Photophysiology. Academic Press, New York, 1: 199-221
3. Bryant DA, Guglielmi G, Tandeau de Marsac N, Castets AM, Cohen-Bazire G (1979) The structure of cyanobacterial phycobilisomes: a model. Arch Microbiol 123: 113-127
4. Cohen-Bazire G, Bryant DA (1982) Phycobilisomes: composition and structure. In: Carr NG, Whitton BA (eds) The Biology of Cyanobacteria. Blackwell, Oxford London, p 143-19o
5. Förster T (1967) Mechanisms of energy transfer. In:Florkin, M, Stotz EH (eds) Comprehensive Biochemistry, Elsevier, Amsterdam London New York 22: 61-8o
6. Füglistaller P, Mimuro M, Suter F, Zuber H (1987) Allophycocyanin complexes of the phycobilisomes from Mastigocladus laminosus. Influence of the linker polypeptide L 8.9 on the spectral properties of the phycobiliprotein subunits. Biol Chem Hoppe-Seyler 368: 353-367
7. Gantt E (1981) Phycobilisomes. Ann Rev Plant Physiol 33: 327-347
8. Gillbro T, Sandström A, Sandström V, Wendler J, Holzwarth AR (1985) Picosecond study of energy transfer kinetics in phycobilisomes of Synechococcus 63o1 and the mutant AN 112. Biochim Biophys Acta 8o8: 52-65
9. Glazer AN (1984) Phycobilisome. A macromolecular complex optimized for light energy transfer. Biochim Biophys Acta 768: 29-51
1o. Glazer AN (1985) Light harvesting by phycobilisomes. Ann Rev Biophys Biophys Chem 14: 47-77

11. Glazer AN, Yeh SW, Webb SP, Clark JH (1985) Disk-to-disk transfer as the rate-limiting step for energy flow in phycobilisomes. Science 227: 419-423
12. Holzwarth AR, Wendler J, Wehrmeyer W (1982) Picosecond time resolved energy transfer in isolated phycobilisomes from Rhodella violacea (Rhodophyceae). Photochem Photobiol 36: 479-487
13. Holzwarth AR (1985) Energy transfer kinetics in phycobilisomes. In: Michel-Beyerle ME (ed) Antennas and Reaction Centers of Photosynthetic Bacteria. Springer, Berlin Heidelberg, p 45-52
14. Holzwarth AR, Bittersmann E, Reuter W, Wehrmeyer W (1989) Studies on chromophore coupling in isolated phycobiliproteins III. Picosecond excited state kinetics and time-resolved spectra of different allophycocyanins from Mastigocladus laminosus. J Biophys (in press)
15. Klotz AV, Glazer AN (1985) Characterization of the bilin attachment sites in R-phycoerythrin. J Biol Chem 26o: 4856-4863
16. Koller KP, Wehrmeyer, Mörschel E (1978) Biliprotein assembly in the disc-shaped phycobilisomes of Rhodella violacea. On the molecular composition of energy-transfering complexes (tripartite units) forming the periphery of the phycobilisome. Eur J Biochem 91: 57-63
17. Lagarias JC, Klotz AV, Dallas JL, Glazer AN, Bishop JE, O'Connel JF, Rapoport H (1988) Exclusive A-ring linkage for singly attached phycocyanobilins and phycoerythrobilins in phycobiliproteins. J Biol Chem 263: 12977-12985
18. Lundell DJ, Yamanaka G, Glazer AN (1981) A terminal energy acceptor of the phycobilisome: The 7o.ooo-dalton polypeptide of Synechococcus 63o1 phycobilisome - A new biliprotein. J Cell Biol 91: 315-319
19. Lundell DJ, Glazer AN (1983) Molecular architecture of a light-harvesting antenna. Structure of the 18 S core-rod subassembly of the Synechococcus 63o1 phycobilisome. J Biol Chem 258: 894-9o1
2o. Mörschel E, Koller KP, Wehrmeyer W, Schneider H (1977) Biliprotein assembly in the disc-shaped phycobilisomes of Rhodella violacea. I. Electron microscopy of phycobilisomes in situ and analysis of their architecture after isolation and negative staining. Cytobiol 16: 118-129
21. Mörschel E, Koller KP, Wehrmeyer W (198o) Biliprotein assembly in the disc-shaped phycobilisomes of Rhodella violacea. Electron microscopical and biochemical analyses of C-phycocyanin and allophycocyanin aggregates. Arch Microbiol 125: 43-51
22. Mörschel E, Wehrmeyer W, Koller KP (198o) Biliprotein assembly in the disc-shaped phycobilisomes of Rhodella violacea. Electron microscopical and biochemical analysis of B-phycoerythrin and B-phycoerythrin-C-phycocyanin aggregates. Eur J Cell Biol 21: 319-327
23. Nies M, Wehrmeyer W (1981) Biliprotein assembly in the hemidiscoidal phycobilisomes of the thermophilic cyanobacterium Mastigocladus laminosus Cohn. Characterization of dissociation products with special reference to the peripheral phycoerythrocyanin-phycocyanin complexes. Arch Microbiol 129: 374-379

24. Ohki K, Gantt E, Lipschultz CA, Ernst MC (1985) Constant phycobilisome size in chromatically adapted cells of the cyanobacterium Tolypothrix tenuis and variation in Nostoc sp. Plant Physiol 79: 943-948
25. Ohki K, Fujita Y (1987) Non-hemidiscoidal phycobilisome in Cyanophyta. In: Biggins J (ed) Progress Photosynth Res 2: 1.157-1.160. Nijhoff/Dr Junk, Dordrecht
26. Padan E (1979) Facultative anoxygenic photosynthesis in in cyanobacteria. Ann Rev Plant Physiol 30: 27-40
27. Redlinger T, Gantt E (1981) Phycobilisome structure of Porphyridium cruentum. Polypeptide composition. Plant Physiol 68: 1375-1379
28. Reuter W, Wehrmeyer W (1988) Core substructure in Mastigocladus laminosus I. Microheterogeneity in two of three allophycocyanin core complexes. Arch Microbiol 150: 534-540
29. Reuter W, Wehrmeyer W (1989) Core substructure in Mastigocladus laminosus II. The central part of the tricylindrical core - AP_{CM} - contains the "anchor" polypeptide and no allophycocyanin B. Arch Microbiol (in press)
30. Schirmer T, Bode W, Huber R, Sidler W, Zuber H (1985) X-ray crystallographic structure of the light-harvesting biliprotein C-phycocyanin from the thermophilic cyanobacterium Mastigocladus laminosus and its resemblance to globin structures. J Mol Biol 184: 257-277
31. Schirmer T, Huber R, Schneider M, Bode W, Miller M, Hackert ML (1986) Crystal structure analysis and refinement at 2.5 Å of hexameric C-phycocyanin from the cyanobacterium Agmenellum quadruplicatum. J Mol Biol 188: 651-676
32. Schirmer T, Bode W, Huber R (1987) Refined three-dimensional structures of two cyanobacterial C-phycocyanins at 2.1 and 2.5 Å resolution. A common principle of phycobilin-protein interaction. J Mol Biol 196: 677-695
33. Schopf JW (1974) Palaeobiology of the Precambrium. The age of blue-green algae. In: Dobshansky T, Hecht MK, Steere WC (eds) Evolutionary Biology 7: 1-43, Plenum Press, New York London
34. Sidler W, Kumpf E, Rüdiger W, Zuber H (1986) The complete amino acid sequence of C-phycoerythrin from the cyanobacterium Fremyella diplosiphon. Biol Chem Hoppe-Seyler 367: 627-642
35. Suter GW, Mazzola P, Wendler J, Holzwarth AR (1984) Fluorescence decay kinetics in phycobilisomes isolated from the blue-green alga Synechococcus 6301. Biochim Biophys Acta 766: 269-276
36. Suter GW, Holzwarth AR (1987) A kinetic model for the energy transfer in phycobilisomes. Biophys. J. 52: 673-683
37. Tandeau de Marsac N, Cohen-Bazire G (1977) Molecular composition of cyanobacterial phycobilisomes. Proc Natl Acad Sci USA 74: 1635-1639
38. Tandeau de Marsac N (1983) Phycobilisomes and complementary chromatic adaptation in cyanobacteria. Bull Inst Pasteur 81: 201-254

39. Wehrmeyer W (1983) Organization and composition of cyanobacterial and rhodophycean phycobilisomes. In: Papageorgiou GC, Packer L (eds) Photosynthetic Prokaryotes. Cell Differentiation and Function. Elsevier/North Holland, New York, p 1-22

4o. Wehrmeyer W (1983) Phycobiliproteins and phycobiliprotein organization in the photosynthetic apparatus of cyanobacteria, red algae and cryptophytes. In: Jensen U, Fairbrothers DE (eds) Proteins and Nucleic Acids in Plant Systematics. Springer, Berlin Heidelberg, p 143-167

41. Wehrmeyer W, Wendler J, Holzwarth AR (1985) Biochemical and functional characterization of a peripheral phycobilisome unit from Porphyridium cruentum. Measurement of picosecond energy transfer kinetics. Eur J Cell Biol 36: 17-23

42. Wehrmeyer W, Zimmermann C, Ohki K, Fujita Y (1988) On a new type of cyanobacterial phycobilisome exemplified in Phormidium persicinum Gomont. Eur J Cell Biol 46: 539-546

43. Wendler J, Holzwarth AR, Wehrmeyer W (1984) Picosecond time-resolved energy transfer in phycobilisomes isolated from the red alga Porphyridium cruentum. Biochim Biophys Acta 765: 58-67

44. Wildman RB, Bowen CC (1974) Phycobilisomes in blue-green algae. J. Bacteriol 117: 866-881

45. Zuber H, Brunisholz R, Sidler W (1987) Structure and function of light-harvesting pigment-protein complexes. In: Amesz J (ed) Photosynthesis. Elsevier, Amsterdam, p 233-271

The molecular architecture of the thylakoid membrane from various classes of eukaryotic algae

C. Wilhelm, I. Wiedemann, P. Krämer, I. Lenartz-Weiler and C. Büchel
Institut für Allgemeine Botanik
Johannes Gutenberg-Universität
Saarstr. 21
D-6500 Mainz
Federal Republic of Germany

Introduction

There is convincing consensus that the photosynthetic apparatus is of prokaryotic origin. The wide variety of algal plastids is mostly assumed to be the result of different endocytological events. Chloroplasts surrounded by two membranes as in rhodophytes and chlorophytes were considered as the association of a prokaryotic symbiont and a eukaryotic host, whereas algae having a chloroplast surrounded with more than two membranes can be delineated from an endocytological event of two eukaryotes (see S. Gibbs in this volume). Since chlorophyll b was neither combined with chlorophyll c nor with phycobiliproteins it was proposed that all present day chloroplasts can be integrated in three lines. Based on recent progress in sequencing the 28 S RNA from various algal classes (Perasso et al., 1989; Adoutte, this volume) it was suggest that these lines emerge as three distinct groups, although within the branches deep demarcations can be observed.

In this contribution the prasinophycean alga Mantoniella squamata will be compared with Chlorella, a typical green alga, and some chlorophyll c-containing algae like Pleurochloris meiringensis (Xanthophyceae) and Prymnesium parvum (Prymnesiophyceae). Mantoniella squamata had been shown to be an exception in chlorophyll b-containing plants. It contains chlorophyll b as well as a chlorophyll c-species (Wilhelm, 1987), the so-called light-harvesting chlorophyll complex (LHC II) deviates with respect to polypeptide structure and pigmentation from other chlorophyll b-bearing proteins (Foss et al., 1985; Fawley et al. 1986). Comparing the molecular organization and function of the thylakoids from Mantoniella with other chlorophyll b- or chlorophyll c-containing systems we adress the question how the green and chromophyte chloroplasts are related phylogenetically.

There are two reasons to concentrate on the biochemical and ultrastructural organization of the LHC and on the functional mechanisms to balance the energy distribution between both photosystems. First, the phylogenetic relevance of the LHC's becomes obvious by the differences in pigmentation, which is still today an important taxonomical criterium. Second, in nearly all chlorophyll c-containing algae the thylakoids are predominantly arranged in threes, whereas in most green algae the thylakoid system is distinctly differentiated in stroma and grana lamellae. In the latter ones the thylakoids are tightly appressed, whereas in chlorophyll c-containing systems the apposition of the membranes is very loose, sometimes with clear gaps between. The LHC II of green plants is thought to be the decisive factor which determines membrane adhesion in the partition regions (Ryrie et al. 1980, Staehelin, 1983). Therefore, alterations in the LHC-polypeptides can change membrane-membrane interaction and, as a consequence of it, the structure and function of the whole thylakoid.

Materials and Methods

Chlorella fusca, Mantoniella squamata, Pleurochloris meiringensis and Prymnesium parvum were grown for biochemical analysis in batch cultures according to Wilhelm et al. (1986). Physiological experiments were done with homocontinuous cultures grown at 20°C in a light/dark cycle of 16/8 h with a light intensity of 1 W/m^2 (LL) or at 20 W/m^2 (HL), respectively. The optical density was adjusted at 4 mg Chl a/l suspension. The isolation procedures and the characterization of the LHC by absorption and fluorescence spectroscopy and pigment analysis were previously reported (Wilhelm et al., 1986; Wilhelm et al., 1989b). The estimation of the average absorption cross section normalized to chlorophyll a and the preparation for electron microscopy was described in Wilhelm et al. (1988a). The measurement of pulse modulated fluorescence induction kinetics was done with a PAM 100 (Fa Waltz, Effeltrich, FRG).

Results

The biochemistry of antenna proteins

The LHC binds at least 50% of the total chlorophylls and 80-90% of the taxonomically relevant xanthophylls. In contrast to the wide variety of light harvesting xanthophylls (about 20) which occur in algal anntennas only three chlorophylls have been found. Chlorophyll a and b in "green algae" and chlorophyll a and c_1/c_2 in "chromophyte algae". In some diatoms and Prymnesiophytes a new chlorophyll c-like pigment, chl c_3, was found (Jeffrey and Wright, 1987; Stauber and Jeffrey, 1988). In addition, in Mantoniella a third chlorophyll, which is extremely simililar to chlorophyll c_1 can be identified (Wilhelm, 1987). All these chlorophylls are chlorophyllides lacking a hydrophobic side chain. Because the fact that different chlorophyllides obligatorily occur as intermediates of the biosynthesis of chlorophyll a, there is urgent necessitiy to define clearly what the term "chlorophyll c" means. Therefore, we propose to define "chlorophyll c" by the following criteria:
Chlorophyll c is defined as Mg-protochlorophyllid-like pigments combining the following features:

1. In vivo it is protein-associated in the light-harvesting complex
2. It functions in energy transfer from carotenoids to chlorophyll a
3. It is a Mg-pheoporphyrin-derivate with an unsaturated ring IV
4. It is not esterified (free acid)
5. Its soret band is 8-10 fold higher than the red absorption maximum.

The chlorophyll c-like pigment in Mantoniella fits with all these criteria. Figure 1 gives the HPLC scan of the pigments isolated from the purified LHC of Mantoniella. The numbers indicate the following pigments: 1: chlorophyll c; 2: neoxanthin; 3: violaxanthin; 4: prasinoxanthin; 5: prasinoxanthin-epoxide; 6: luteoxanthin; 7: chlorophyll b; 8: chlorophyll a; 9: carotenes. Figure 2 shows the absorption spectrum of the fast running fraction. The function in energy transfer to chlorophyll a is demonstrated in figure 3. Exciting the LHC with a wavelength of 465 nm for chlorophyll c (fig 3a) the low temperature emission spectrum exhibits only chlorophyll a emission, due to an efficient energy transport from chlorophyll c to chlorophyll a. In contrast, when the pigments are mildly deconnected, a chlorophyll c emission at 636 nm will appear (fig 3b). Scanning the excitation spectrum at an emission wavelength of 636 nm, a clear chlorophyll c-in vivo spectrum is the result (fig.4). Therefore,it can be concluded that chlorophyll c is a true light harvesting pigment in the LHC of Mantoniella.

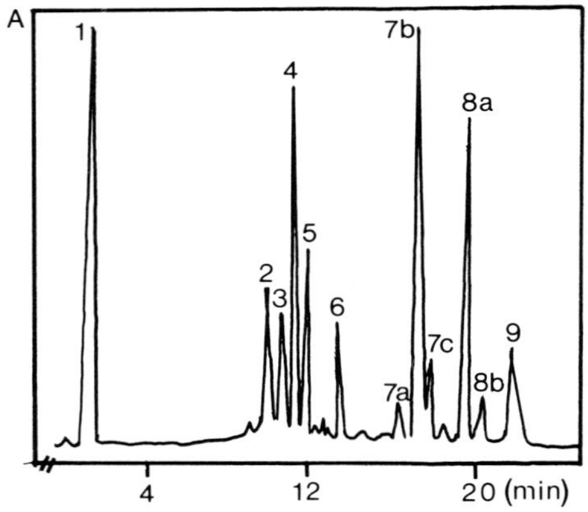

Fig. 1 HPLC-scan of the pigments isolated from LHC of Mantoniella squamata; for numbers see text.

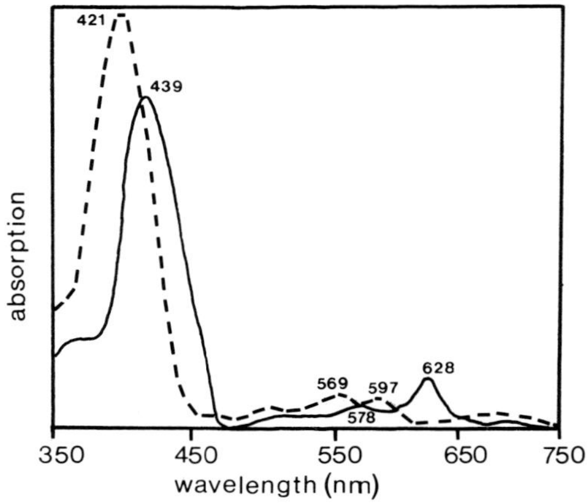

Fig. 2 Absorption spectrum of chlorophyll c (solid line) and its pheopigment (dotted line) isolated from Mantoniella squamata.

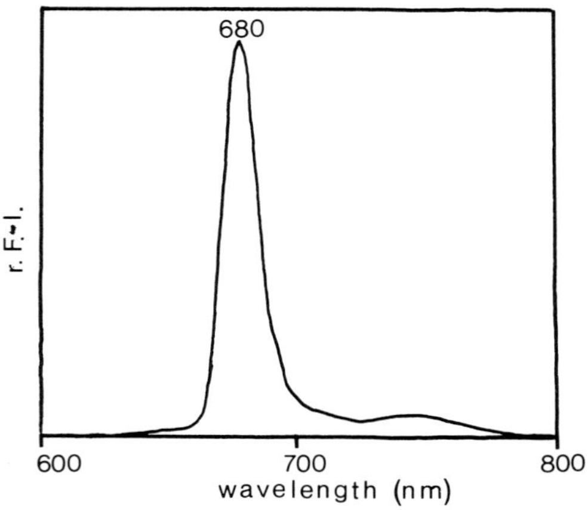

Fig. 3 Fluorescence emission spectrum of the LHC of Mantoniella at 77K. (Excit:465nm)

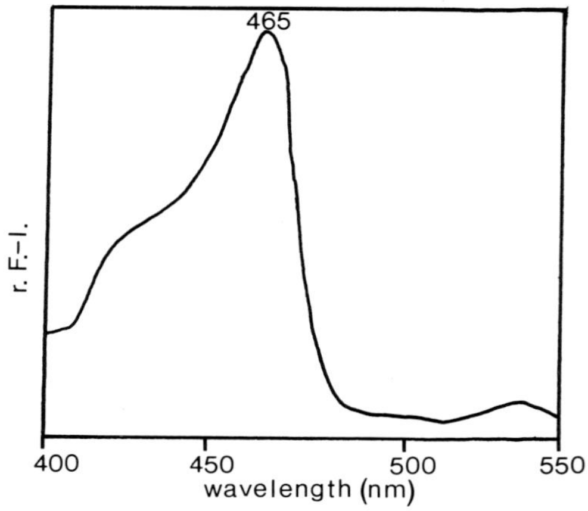

Fig.4 Fluorescence excitation spectrum of mildly deconneced LHC of Mantoniella at 77K (Em: 636 nm)

In chromophytes the xanthophylls are assumed to be the predominating photosynthetic pigments, whereas in green algae the chlorophylls are the major ones. In the last few years the molar ratios xanthophyll per chlorophyll had been measured. In table 1 this ratio is compared in four different species on the level of total cells as well as on the level of the isolated light antenna proteins. Obviously, Prymnesium binds ten times more xanthophylls per chlorophyll than Chlorella, whereas the xanthophyte and the prasinophyte hold an intermediate position. It has to be emphasized that these differences also occur on the level of the isolated protein, and differences in the pigment loading of the polypeptides are very likely. Therefore, the ratio pigments per polypeptide were analyzed. The results for Chlorella and Mantoniella are given in table 2.

Table 1
The ratio xanthophyll per chlorophyll a (mol xan/1000 mol Chl a) in Chlorella fusca, Mantoniella squamata, Pleurochloris meiringensis and Prymnesium parvum.

	Chlorella	Mantoniella	Pleurochloris	Prymnesium
cells	184	420	220	1240
LHC	258	521	510	-

Table 2.
Pigments per polypeptide in Chlorella fusca and Mantoniella squamata

	Chlorella	Mantoniella
chl a	8	15
chl b	7	17
xan	2-3	7-8
chl c	-	2-3
neoxanthin	0,5	1-2
trihydroxy-α-carotene	1	-
violaxanthin	0,5	1
lutein	1	-
prasinoxanthin	-	4-5

The LHC from Mantoniella does not only bind 2 times more xanthophylls per chlorophyll a, but also its polypeptides are associated with double amounts of chlorophylls. In total, the LHC of Chlorella binds 18 pigments molecules per one polypeptide, whereas in Mantoniella this ratio is 40. This is a good indication that the polypeptide structure, namely the pigment binding sites, are strongly modified in Mantoniella in comparison to other green algae. The binding sites for the pigments are thought to be in the alpha-helices in the three membrane spanning residues (Thornber et al.1988). In the reaction center proteins the chlorophylls are presumably associated to histidines, which is not very likely in the LHC, because its alpha-helices have a paucity in histidines. The involvement of glutamine and aspraragine residues in addition to histidine seems a good possibility, which could also explain a hypothetical multiplicity for binding sites.

In addition, the pigment pattern of the isolated LHC is not as fixed as it has been considered. Figure 5 gives evidence that the LHC from Mantoniella changes it composition in response to light intensity during growth. In HL an additional pigment will be present whose identity is not yet determined. There is some evidence that it is a new carotenoid (Breton, pers. com.).

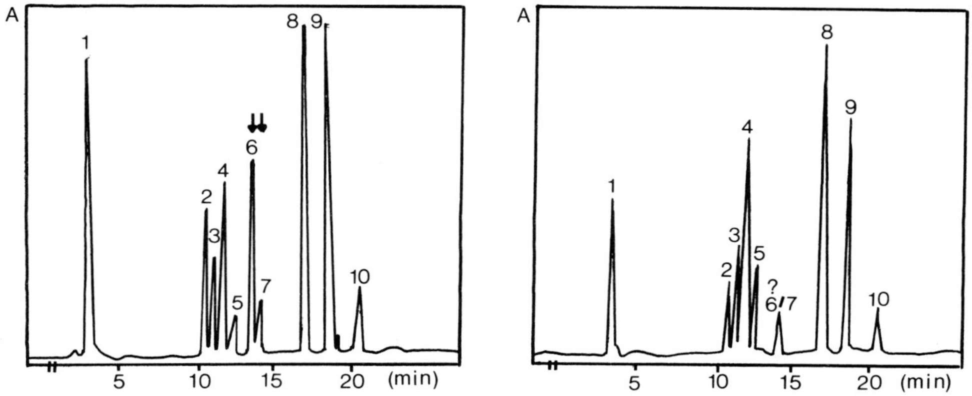

Fig.5 HPLC-scan of the pigments isolated from LHC of Mantoniella grown under 20 W/m² (left hand) and under 1 W/m² (right hand). 1: Chl c; 2: neo.; 3: viola.; 4: prasino.; 5: prasino-ep.; 6: unknown component (see arrow); 7: luteoxanthin; 8: chl b; 9: chl a; 10: carotene.

It is consensus that the chl b-binding algal antenna proteins are very similar to those of higher plants with respect to size of the subunits and their immulogical distances. Therefore, the immunological cross reactivity of the different LHC's was tested (table 3). In chl c-containing LHC's in all cases the subunits are smaller in size than in green algae, but the immunological relationships are very controversial. In some cases the immunological response to chl b-binding peptides is positive in others it is not. Depending on the specificity and the numbers of the epitopes, which the antibody can recognize, different results will be obtained. Therefore, the immunological cross reactivity does not reveal in all cases true phylogenetic distances. Nevertheless, the fact that no antibody is able to recognize the LHC of Mantoniella strengthens its atypical structure.

Table 3
Molecular weights and immunological cross reactivity of the major LHC's of representatives of various algal clases.

	Chlorella	Mantoniella	Pleurochloris	Prymnesium
subunits (kDa)	25; 35	23	23	19; 25

<u>Immunological cross reactivity:</u>

antibody of: with protein of	higher plant	Pedinomonas	Fucus	Pavlova	Phaeodactylum
higher plant		+ (1)		+ (2)	
Chlamydomonas	+ (3)				– (4)
Mantoniella		– (1)	– (5)		
Fucus	+ (6)				
Phaeodactylum				+ (2)	+ (4)
Pleurochloris			– (5)		
Croomonas				– (2)	– (4)
Amphidinium				– (2)	– (4)

<u>The physiology of antenna proteins</u>

The physiological role of LHC's is to absorb the photosynthetic radiation and to funnel the energy as efficiently as possible to the photochemical active centers. The optimization of this function needs a powerful regulation of the energy distribution from the antennae to the photosystems after the need of their activities in relation to the electron flow.

The concept of the lateral heterogeneity of the photosynthetic membrane (Staehelin, 1983), although generally accepted, has provoked some intriguing questions about the regulation of the spatially separated photosystems and antennas. In higher plant chloroplasts the wavelength-dependent

1 Fawley et al. 1986
2 Hiller at al. 1988
3 Chua and Blomberg, 1979
4 Friedman and Alberte, 1986
5 Berkaloff pers. communication
6 Caron et al. 1988

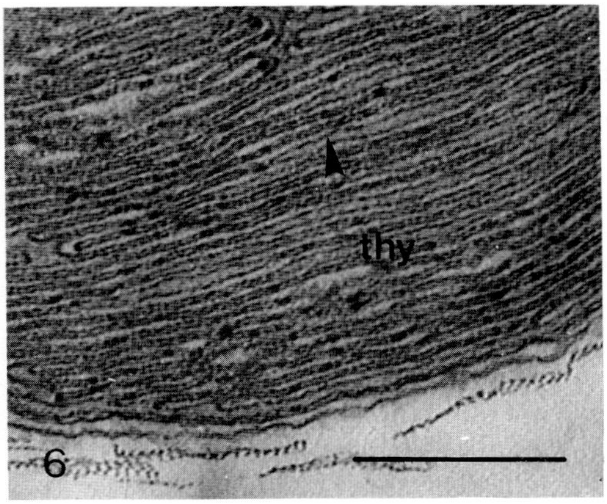

Fig. 6 The ultrastructure of the thylakoids of Mantoniella The thylakoids (thy) are running in parallel with clear gaps between (arrow). bar: 0,25 µm

Tab. 5 The in-vivo absorption cross section per chlorophyll a (k_c) of low and high light grown cells of Mantoniella and Dunaliella.

species	light intensity during growth	k_c-value	reference
Mantoniella	1 W/m^2	0,0071	this paper
	20 W/m^2	0,0074	this paper
Dunaliella	14 W/m^2	0,047	Berner et al. (1989)
	140 W/m^2	0,088	Berner et al. (1989)

in vivo absorption. It can be concluded from these experiments that the thylkoid system in Mantoniella is organized like a pigment bed with a relatively high homogeneity of the distribution of the membrane-intrinsic proteins. This suggestion fits very well with the observation, that state 1-state 2 transitions does not occur. Therefore, it was looked for other types of regulations which balance the activity of both photosystems and regulate the electron flow between in response to changing light climates.

Figure 7a gives the measurement of F_{max} of dark adapted Mantoniella cells. Under conditions when the pulsed light is not actinic and therefore no pH-gradient can be formed the addition of an uncoupler does not influence F_{max}. Therefore, we have to assume that the thylakoid membrane is energized even in the dark. This is in good agreement with an earlier observation (Wilhelm et al. 1989b) that in Mantoniella the fluorescence induction curve

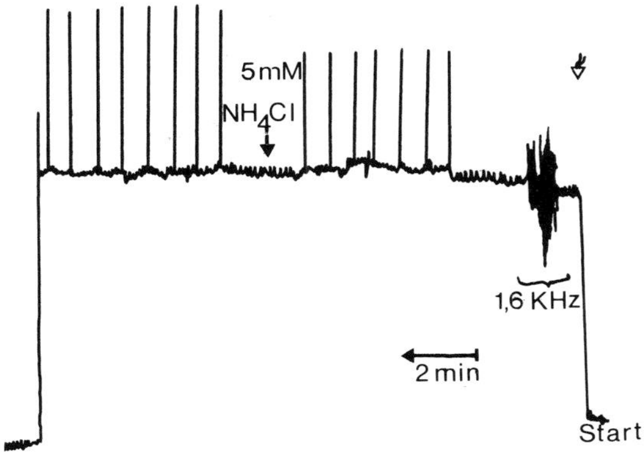

Fig. 7a
Fluorescence induction curve of dark of dark adapted Mantoniella cells.

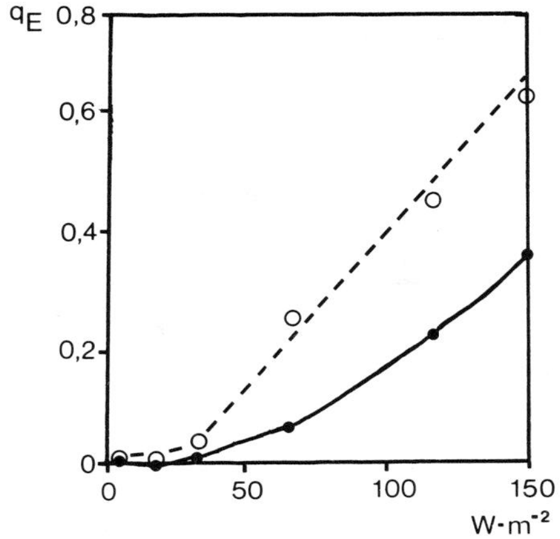

Fig. 7b
The non-photochemical quenching (q_E) in Pleurochloris in relation to the intensity of preillumination.

of dark adapted cells does not show light-induced non-photochemical quenching. It was suggested that the origin of the membrane energization is a chlororespiratorian activity (Wilhelm and Duval, 1989). In the dark the electrons flow from NADPH via a membrane bound NADPH dependent plastochinone-oxidoreductase to plastoquinone and migrate by a cytochrome c-oxidase to oxygen, ultimately. This pathway creates a pH-gradient in the dark and is thought to be suppressed in the light by the competetive activity of the light driven electron flow. In Pleurochloris the situation is found to be very similar, but the chlororespiration is strictly light-regulated (fig 7b). When the cells are preadapted to light, the chlororespirations will be switched off and q_E, which is due to the pH-gradient, becomes visible after a short period of dark incubation. The higher the preillumination the higher is the value of q_E found after the dark period. The light intensity needed to turn off chlororespiration is different in high light and low light grown cells indicating that the chlororespiratorian capacity is of ecophysiological importance.

It can be shown that a high chlororespiratorian activity is strictly correlated with a very strong cooperativity between LHC and PS II, whereas the enhancement of PS I antenna size is observed only, when chlororespiration is decreased. Therefore, chlororespiration is thought to control the energy state 1-state 2-transitions had been intensively analyzed (Fork and Satoh, 1986; Williams and Allen, 1987). The most current model assumes that the overexcitation of PS I (or dark adaption) leads to state 1, that is an increase in the fraction of excitation energy which is utilized by PS II. Selective excitation of PS II is thought to induce the reverse regulation. The molecular basis of state 1-state 2 transitions was explained by reversible LHC IIb phosphorylation by a kinase-system. The activity of this LHC -specific kinase is believed to be controlled by the redox state of the plastoquinone pool. This carrier system is sensible to detect any imbalance in the distribution of excitation energy. An overexcitation of PS II leads to an reduced PQ pool which activates the LHC II kinase. The phosphorylated LHC is thought to become deconnected from PS II and will migrate out of the partition regions.

Briantais et al. (1984) showed that the membrane distances are of very high physiological importance. They demonstrated that the cooperation of the PS II/LHC units is different under stacked and unstacked conditions. In destacked thylakoids the variable fluorescence is lowered, indicating that the PS II activity is decreased by less effective energy transfer from the LHC. Therefore, the morphological observation of appressed membranes has a physiological or functional equivalent. Summarizing it must be emphazised that the physiological activity of PS II depends on its cooperation with the LHC II, which is regulated by the membrane-membrane interaction in the grana regions.

It is known so far from electron microscopical studies that green algae form thylakoids which are clearly differentiated in exposed and appressed regions. In chlorophyll c-containing chloroplasts, however, the apposition of the lamellae is somehow looser. The question arises if this observation is due to the preparation conditions or if it reflects a biologically relevant difference. Indeed, the distances between the membranes can be altered very easily during the long parcours of EM-fixation and though it remains doubt about the physiological relevance of this result.

Nevertheless, wavelength-dependent state 1-state 2 transitions had been found to be lacking in most of chl c-containing algae, which had been analyzed. Although LHC-phosphorylation had been observed (Caron et al., 1988), changes in the pattern of energy distribution became obvious not after preillumination with different wavelengths, but after applying different light intensities (Owens, 1986; Büchel et al., 1988). In Mantoniella, however, no changes in energy distribution had been found (Wilhelm et al. 1989b). As shown in table 4 the ratio F_{max}/F_o does not depend on the light quantity or quality of preillumination:

Table 4
The ratio of F_{max}/F_o in Mantoniella squamata after different preilluminations (20 min) before measurement.

preillumination condition	F_{max}/F_o	number of experiments
dark adapted	1,4 ± 0,15	8
710 nm (21,5 µE m^{-2} sec^{-1})	1,5 ± 0,1	8
520 nm (29,4 µE m^{-2} sec^{-1})	1.4 ± 0,15	7
470 nm (32,5 µE m^{-2} sec^{-1})	1,45± 0,1	10
440 nm (34,7 µE m^{-2} sec^{-1})	1,5 ± 0,05	9
(white light) 50 µE m^{-2} sec^{-1}	1,6 ± 0,1	6
(white light) 100 µE m^{-2} sec^{-1}	1,7 ± 0,15	8
(white light) 200 µE m^{-2} sec^{-1}	1,6 ± 0,1	7

The ultrastructure of the thylakoids from Mantoniella is shown in figure 6. The thylakoids are running in parallel neither forming true partitions regions nor organized in threes. Between the lamellae in most cases a clear gap can be seen, indicating that membrane-membrane interactions are not very likely. Therefore, there is good agreement between the fixed ratios of F_{max}/F_o after different preilluminations and the ultrastructural arrange ment of the lamellae. In order to elaborate further evidence, that in Mantoniella the thylakoids are functioning like single membranes, the cells were grown at different light intensities. In typical green algae the thylakoids form more appressed membranes relative to unappressed ones when grown under low light intensities. Dubinsky et al. (1986) showed that stacked membranes exhibit a more pronounced so-called "package effect". It means that the in-vivo absorption cross section per chlorophyll a (k_c) is decreased. Table 5 gives the k_c-values of low light and high light adapted cells of Mantoniella and of Dunaliella reported by Berner et al. (1989).

The observation that the k_c-value of Mantoniella is tenfold lower than in Dunaliella is due to the fact that its chloroplast is much smaller in size and the stacking does not contribute efficiently to the enhancement of the distribution from LHC to the photosystems. In contrast to the wavelength-dependent state 1-state 2 transitions in green algae which are triggered by the redox level of the PQ-pool and/or the pH-gradient, in Pleurochloris these parameters are kept constant and the distribution of the energy seems to be regulated by the ATP/ADP and/or NADPH/$NADP^+$ ratios. This consideration explains the paradoxon that Pleurochloris does not exhibit wavelenght-dependent but light intensity-dependent state transitions.

Discussion

1. The phylogenetic implications of chlorophylls

The traditional view that the algae can be grouped in three lines, the chlorophyll c-containing, so-called "chromophytes", the chlorophyll b-containing chlorophytes and the phycobiliprotein containing algae must be modified. There is convincing evidence that the cytological and biochemical features of the host cell are of higher taxonomical importance than those of the symbiont. Second, although the "chlorophytes" on the one hand and the "chromophytes" on the other hand share a lot of common cell features, a more detailed analysis gives evidence that there are very long phylogenetic distances within the groups when comparing "primitive" with advanced forms. Third, chlorophyll c has to be considered as a very doubtful taxonomic criterium. Assuming that biosynthetical pathways reflect phylogenetic roots only these constituents can be used as phylogenetic markers of "advanced" taxa which are the products of late branchings in biosynthesis. Chlorophyll c, however, is a derivate of a chlorophyll a precursor, which may be virtually present in all oxygen evolving organisms. Therefore, chlorophyll c is not an indicator for a specialized group but a good hint for primitiveness.

2. The phylogenetic implications of membrane stacking

The differention of grana implies a spatial separation of light-harvesting and water splitting at the one side and $NADP^+$-reduction and ATP synthesis on the other side. The balance of both processes in response to changing light climates or nutrition conditions demands for efficient regulation mechanisms. It is very likely to assume that the higher plant type is the most advanced one, whereas systems with less or no lateral heterogeneity are more primitive. Using membrane-membrane interactions as a phylogenetic criterium, an evolution line may be postulated starting from totally unstacked systems - like blue-green algae or Mantoniella - to systems with increased membrane adhesion - perhaps most of the "chromophytes" - and ending with the highly spatially differentiated grana-type chloroplasts.

3. The phylogenetic implications of chlororespiration

Keeping in mind, that the thylkoid is a derivate of the procaryotic cytoplasma membrane it is very likely to assume, that in some species the respiratorian competence had been conserved. This primitiveness can be observed in blue-green algae as well in those eukaryotic algae which possess full chlororespiration. In this case the photosynthetic membrane is a permanent source for ATP-synthesis and the light is used mainly to replace dark respiration and when ATP supply is saturated CO_2-assimilation is activated. In such a system there are different sinks and sources of electrons in the thylakoid and the activity of both photosystems can be simply regulated by competition. It can be speculated that by the loss of the cytochrome c-oxidase as important sink and the NADPH-dependent PQ-oxidoreductase as source the PQ-pool becomes sensitive to changing light climates and therefore could be used as a trigger for a special energy distribution mechnism. The chlorophyll c-containing algae may contain a wide variety of these both extremes and therefore very unusual energy distribution pattern may be found.

Acknowledgements

This work was supported by the "Deutsche Forschungsgemeinschaft" (Grant Wi 764/1-4) and by the "Deutscher Akademischer Austauschdienst" for scientific exchange with the Ecole Normale Superieure, Paris.

References

Berner T, Dubinsky Z, Wyman K, Falkowsky PG (1989) Photoadaptation and the package effect in Dunaliella tertiolecta. J Phycol 25:70-78

Briantais JM, Vernotte C, Olive J, Wollman FA (1984) Kinetics of cation-induced chnages of photosystem II fluorescence and of lateral distribution of the two photosystems in the thylakoid membranes of pea chloroplasts. Biochim Biophys Acta 766:1-8

Büchel C, Wilhelm C,Lenartz-Weiler I (1988) The molecular analysis of the light adaptation reactions in the yellow-green alga Pleurochloris meiringensis (Xanthophyceae). Bot Acta 4: 306-310.

Caron L, Remy R, Berkaloff C (1988) Polypeptide composition of light-harvesting complexes from some brown algae and diatoms. FEBS Letters 229:11-15.

Chua N, Blomberg F (1979) Immunochemical studies of thylakoids membrane polypeptides from spinach and Chlamydomonas reinhardii. Modified procedure for crossed immunoelectrophoresis of dodecyl sulfate protein complexes.J Biol Chem 254:215-223.

Dubinsky Z, Falkowsky PG Wyman K (1986) Light harvesting and utilitzation by phytoplankton. Plant Cell Physiol 27:1335-1349

Fawley MW, Stewart KD, Mattox KR (1986) The novel light-harvesting pigment-protein complex of Mantoniella squamata (Chlorophyta). J Mol Evol 23:168-170

Fork DC, Satoh K (1986) The control by state transitions of the distribution of excitation energy in photosynthesis. Annu Rev Plant Physiol 37: 335-361

Foss P, Guillard R, Liaaen-Jensen S (1985) Carotenoids from eukaryotic ultraplancton clones (Prasinophyceae). Phytochemistry 25: 119-124

Hiller R, Larkum AWD, Wrench P (1988) Chlorophyll-proteins of the prymnesiophyte Pavlova lutherii (Droop) comb. nov.: Identification of the major light-harvesting complex. Biochim Biophys Acta 932:223-231.

Jeffrey SW, Wright SW (1987): A new spectrally distinct component in preparations of chlorophyll c from the micro-alga Emiliania huxleyi (Prymnesiophyceae). Biochim Biophys Acta 894: 180-188.

Owens TG (1986) Photosystem II heterogeneity in the marine diatom Phaeodactylum tricornutum. Photochem Photobiol 43:535-544.

Perasso R, Baroin A, Qu LH, Bachellerie JP, Adoutte A (1989) Origin of the algae. Nature 339: 142-144

Ryrie I, Anderson J, Goodschild D (1980) The role of light-harvesting chlorophyll a/b-protein complex in chloroplast membrane stacking. Eur J Biochem 107: 345-354

Staehelin LA, Arntzen CJ. (1883): Regulation of chloroplast membrane function. J Cell Biol 97: 1327-1337

Thornber JP, Peter GF, Chitnis PR, Nechustai R, Vainstein A (1988) The light-harvesting complex of photosystem II of higher plants. In: Stevens SE, Bryant DA (eds.) Light energy transduction in Photosynthesis, The American Society of Plant Physiologists, p 137.

Stauber J, Jeffrey S (1988) Photosynthetic pigments in fifty-one species of marine diatoms.J Phycol 24:158-172.

Wilhelm C, Duval JC (1989) Fluorescence induction kinetics as a tool to detect chlororespiratorian activity in the prasinophycean alga Mantoniella squamata. Biochim Biophys Acta. (in press).

Wilhelm C, Lenartz-Weiler I, Wiedemann I, Wild, A (1986) The light harvesting system of Micromonas species (Prasinophyceae): The combination of three different chlorophyll species in one single chlorophyll protein complex. Phycologia 25:304-312.

Wilhelm C (1987) Purification and identification of chlorophyll c-1 from the green alga Mantoniella squamata Biochim Biopyhs Acta 892:23-29.

Wilhelm C, Krämer P, Lenartz-Weiler I (1989a) A new molecular strategy of the energy distribution between the photosystems by extreme variation in photosystem I/II stoichiometries probed in the chlorophyll b-containng alga Mantoniella squamata. Photosynth Res 20:221-233

Wilhelm C, Wiedemann I, May M (1989b) Comparative analysis of the composition of two chlorophyll b containing light harvesting complexes. Planta in press.

Williams WP, Allen JF (1987) State 1-state 2 changes in higher plants and algae. Photosynth Res 13: 19-45.

Chlororespiration: A respiratory activity in the thylakoid membrane of microalgae and higher plants

A. Vermeglio, J. Ravenel and G. Peltier

Service de Radioagronomie

CEN CADARACHE

F-13108 SAINT PAUL-LEZ-DURANCE CEDEX

FRANCE

INTRODUCTION

In living cells respiration and photosynthesis are the two major bioenergetic processes which supply the energy required for growth and reproduction. Both activities are present in the majority of photosynthetic organisms: purple bacteria, cyanobacteria, algae and higher plants. During respiration the free energy liberated by the oxidation of reduced compounds by molecular oxygen is partially utilized for phosphorylation of ADP into ATP. During oxygenetic photosynthesis, light energy conversion induces the formation of strong reductants with synthesis of ATP from ADP and Pi at the expense of water photooxidation. Both processes necessitate the opera-

----

Abbreviations

CCCP: carbonyl cyanide m-chlorophenylhydrazone.

DBMIB: 2-5 dibromo-3-methyl-6-isopropyl parabenzoquinone.

DCCD: N-N'dicyclo-hexyl carbodiimide.

DCMU: 3-(3,4-dichlorophenyl)-1,2-dimethylurea.

SDS: sodium dodecylsulfate.

SHAM: salicyl hydroxamic acid.

tion of different electron transfer components inserted in a close membrane vesicle (Figs. 1 and 2). Most photosynthetic procaryotes develop both photosynthetic and respiratory electron transport chains in the same membrane: the intracytoplasmatic membrane in the case of photosynthetic bacteria and the thylakoid membrane for cyanobacteria. Several electron carriers are common to both chains: quinones, cytochrome bc complex, soluble cytochrome etc...(Fig. 1).

In the case of purple photosynthetic bacteria, clear-cut evidence for interaction between these two electron transfer chains is the total reversible inhibition of respiration observed under continuous illumination (Van Niel, 1941; Nakamura, 1937). This light-induced inhibition is suppressed by the addition of uncouplers like CCCP. Two mechanisms are responsible for this phenomenon. The first is a thermodynamic control by the light-induced membrane potential at the level of the two first coupling sites of the respiratory chain (McCarthy and Ferguson, 1982; Cotton et al., 1983). This type of inhibition is prevalent under continuous illumination. The second type of this mechanism involves light-induced-changes in the redox state of electron carriers which are common to both respiratory and photosynthetic activities (Richaud et al., 1986). The similarities between the kinetics of respiratory inhibition and those of the light-induced redox change of cytochromes c_1+c is best explained in a model where the respiratory and photosynthetic pathway are linked through the agency of the ubiquinone pool and not the cytochrome c (Lavorel et al., 1989) (Fig. 1).

In the literature on cyanobacteria there is much evidence for an interaction between respiratory and photosynthetic electron transport chains (see Scherer et al., 1988 for review). In several species of cyanobacteria both carbohydrate catabolism and respiratory oxygen uptake are inhibited

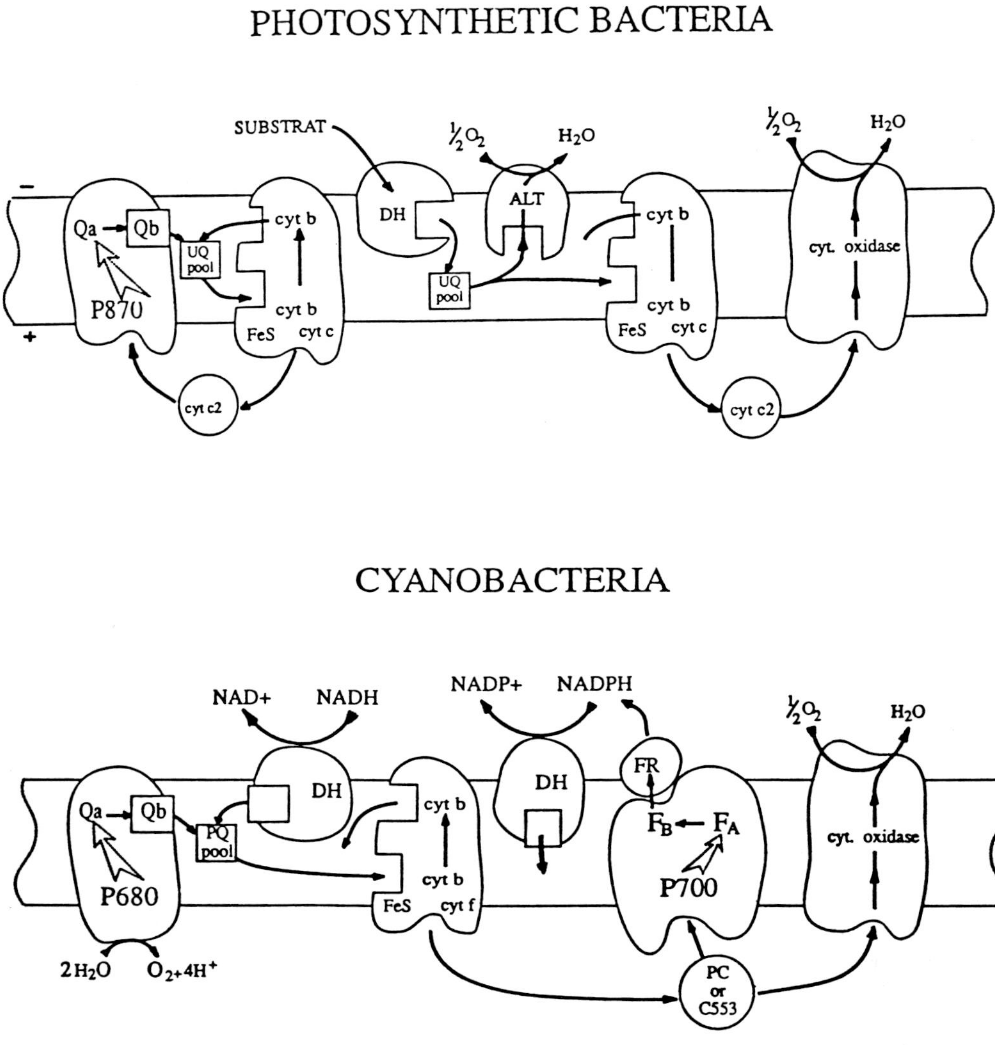

Fig. 1: Schematic representation of the interaction between photosynthetic and respiratory electron components in photosynthetic bacteria and cyanobacteria. DH: dehydrogenase; FR: ferredoxin oxidoreductase; F: ferredoxin; ALT: alternative oxidase; PC: plastocyanin. Light reactions are symbolized by the large arrows.

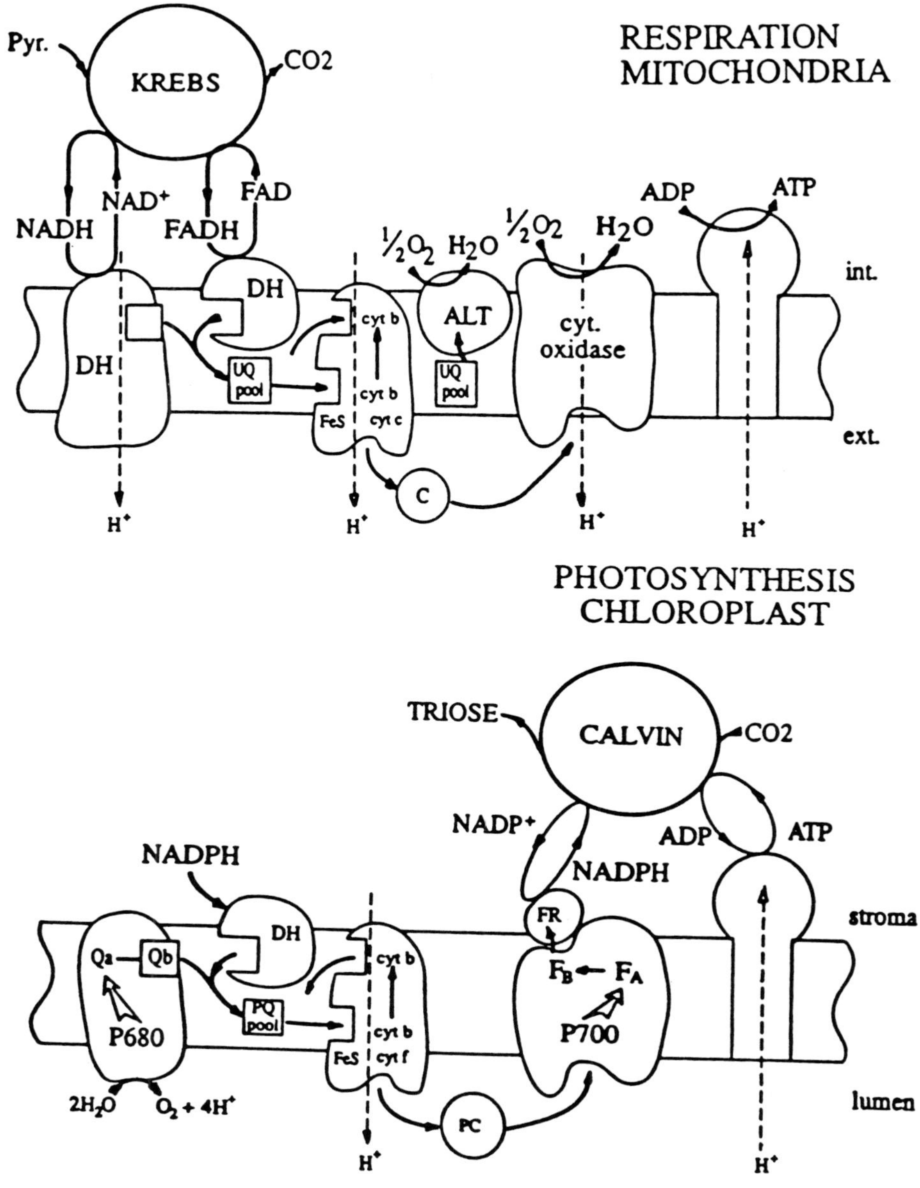

Fig. 2: Schematic representation of components involved in the photosynthetic (chloroplast) and respiratory (mitochondrial) activities.

(50 to 85 %) by light. A dual function of plastoquinone, cytochrome b6f complex and cytochrome C553 on both respiratory and photosynthetic electron transport chains has been demonstrated by the influence of specific inhibitors of these complexes on the respiratory activity (Binder et al., 1984; Matthijs et al., 1984; Hirano et al., 1980; Lockau, 1981). The mechanism of light inhibition appears to be due to competition for electrons between photosynthetic and respiratory chains at the level of photosystem I and cytochrome oxidase (Fig. 1). However, inhibition of respiration in cyanobacteria could not be demonstrated (Scherer et al., 1988).

In photosynthetic eukaryotic organisms, respiration and photosynthesis are believed to proceed in two different organelles: mitochondria and chloroplasts (Fig. 2). In the present article we will review the experimental evidence for a chloroplastic respiration activity in eukaryotes.

I. THE ENTRY OF ELECTRONS INTO THE CHLOROPLAST TRANSPORT CHAIN

Hydrogen evolution can be observed in the light after adaption of microalgae under anaerobic conditions (Gaffron and Rubin, 1942). Since photoevolution of hydrogen is sensitive to DCMU, an inhibitor of photosystem II, a source of electrons different to that from the oxidation of water is likely involved in this process (Kaltwasser et al., 1969; Healey, 1970a,b; Gfeller and Gibbs, 1984). This electron donor has been assumed to be formed during the breakdown of carbohydrate via the citric acid and the glyoxylate cycles (Healey 1970a, b; Stuart and Gaffron, 1971; Gibbs et al., 1986). The sensitivity of acetate dependent hydrogen photoproduction to DBMIB, a plastoquinone antagonist (Böhme et al., 1971), suggests a central role for chloroplastic plastoquinone in this redox reaction process (Gfeller and Gibbs, 1985).

Evidence for chloroplastic NADH and succinic dehydrogenase arrives from several observations. In cell-free extracts, NADH can donate electrons to the photosynthetic electron transport chain in order to photoevolve hydrogen (Ben-Amotz and Gibbs, 1975). Moreover, Godde and Trebst (1980) have shown that a chloroplast fraction from Chlamydomonas reinhardtii can oxidize NADH in the light. This reaction is sensitive to rotenone, DBMIB and DCCD which suggests the occurrence of a rotenone sensitive NADH dehydrogenase coupled to the plastoquinone reduction site of the photosynthetic transport chain. In addition, Peltier and Thibault (1988) have observed low rates of oxygen evolution in a mutant devoid of photosystem I activity. The sensitivity of this oxygen evolution to CCCP (Peltier and Thibault, 1988) and rotenone (Ravenel and Peltier, unpublished results) leads to the conclusion that oxidation of water induces a reduction of NAD(P) via the NAD(P)H-plastoquinone-oxidoreductase using the light-induced membrane potential. The reversibility of the NADH oxidoreductase has already been observed in photosynthetic bacteria (see Knaff, 1978).

Very recently, Willefort et al. (1989) have shown, using a preparation of intact chloroplasts from Chlamydomonas reinhardtii which were devoid of mitochondria and cytoplasmic contamination, that 25 % of the cellular succinate dehydrogenase activity was located within the chloroplast. Malonate is a competitive inhibitor.

Chlorophyll fluorescence-yield measurements,a good indicator of the redox state of the PQ pool, has also been used to demonstrate dark electron input in the chloroplast electron transfer chain. As early as 1963, Goedheer postulated a dark regulation of the redox state of the PQ pool from the analysis of fluorescence transient in Chlorella cells. More recently Bennoun (1982) has shown by fluorescence induction that addition of NADH or NADPH to cell homogenates of *Chlamydomonas* reduced the PQ pool in the dark.

II. THE CHLORORESPIRATORY PLASTOQUINONE OXIDATION PATHWAY

Little information concerning chloroplastic oxidase(s) can be found in the literature. From fluorescence yield measurements Bennoun (1982) has shown that addition of several oxidase inhibitors in the dark induces a large reduction of the chloroplastic PQ pool (Fig. 3). In *Chlamydomonas reinhardtii*,the dark oxidation of the PQ pool is inhibited by CO, NO, KCN, whereas SHAM has no effect. In *Chlorella pyrenoidosa*, addition of KCN or CO does not affect the redox state of plastoquinone while addition of SHAM induces its reduction (Bennoun, 1982). This was taken as evidence for the existence of an electron transport pathway from plastoquinone to oxygen in the thylakoid membrane. This respiration-like process was termed chlororespiration by Bennoun (1982). Caron et al. (1987) observed a quenching of fluorescence in the presence of DCMU in *Phaeodactylum tricornutum* which was cancelled by addition of NH Cl. Quenching was attributed to a pH-induced chlororespiration activity. The interpretation of the fluorescence measurements are, however, ambiguous since the fluorescence yield can be modulated by the redox state of the PQ pool, but also by the membrane energization or the state 1 state 2 transitions. Moreover, the PQ pool reduction can be inhibited by the electron output. Rebeille and Gans (1988) have shown that acceleration of glycolysis either by specifically inhibiting mitochondrial respiration or by addition of uncouplers, induced a large reduction of the PQ pool in the dark.

To obtain direct information on the oxidase(s) characteristics, Peltier et al. (1987) have studied by amperometric and mass spectrometric methods

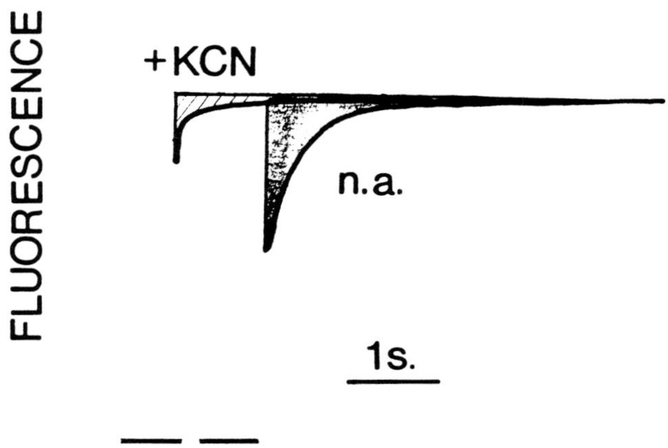

Fig. 3: Chlorophyll fluorescence kinetics for <u>*Chlamydomonas*</u> *mutant (F15/4+) lacking photosystem I activity. The area over the induction curves are a measure of the oxidized plastoquinone molecules present prior to illumination. Addition of 2 mM KCN strongly reduces the PQ pool.*

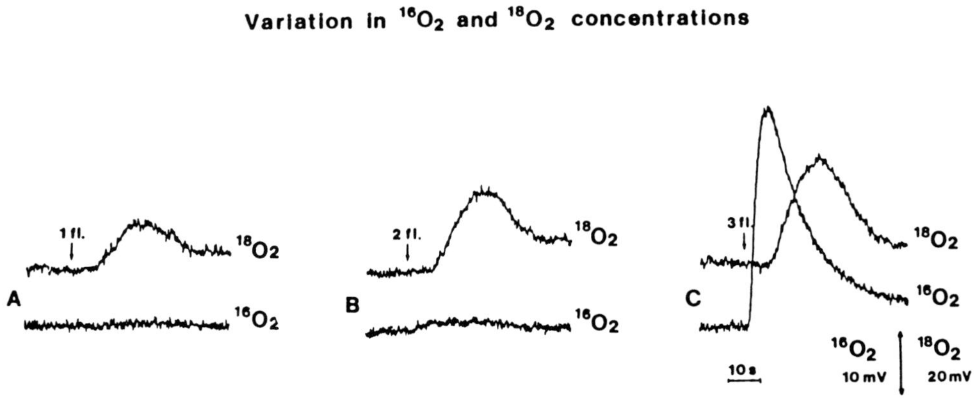

Fig. 4: Mass spectrometer measurement of mass 32 and 36 following the excitation of dark adapted <u>*Chlamydomonas*</u> *cells by one (A). two (B) or three (C) xenon flashes spaced by 0.1 s. The $^{16}O_2$ was removed by bubbling N_2 , and $^{18}O_2$ (98 % $^{18}O_2$) was injected in the gas phase. Under these conditions, changes in mass 32 reflect the $^{16}O_2$ evolved from photolysis of $H_2\,^{16}O_2$ and mass 36, the changes in $^{18}O_2$ linked to respiratory activity.*

the flash induced oxygen exchange in Chlamydomonas whole cells. These authors observed (Fig.5B) an increase of the amperometric signal after the two first flashes. Mass spectrometer experiments performed in the presence of $^{18}O_2$ showed that this signal was not due to the photolysis of water ($H_2^{16}O$) (Fig. 4). The insensitivity of the signal to 10 µM DCMU (Fig. 5B), its stimulation by 1 mM acetate or high oxygen concentration as well as its inhibition by 0.5 mM cyanide indicated that these flash-induced changes in oxygen concentration were related to the inhibition of a respiratory process. Because this inhibition of respiration is insensitive to simultaneous addition of inhibitors of mitochondrial respiration (antimycin A at low concentration and SHAM), and because it occurs on a single flash illumination, these authors concluded that the respiratory activity responsible takes place inside the chloroplast.

The chloroplast oxidase possesses a high Km(O_2) (about 23 µM) compared with that measured for the mitochondrial reactions (0.2 µM for the cytochrome oxidase pathway and 5.5 µM for the alternative pathway; Peltier et al., 1987). The effect of different inhibitors allows to differentiate between chlororespiration and mitorespiration. As already stated, chlororespiration is sensitive to a relative low concentration of KCN (0.5 mM) which does not affect the mitorespiration. On the other hand, addition of antimycin A plus SHAM inhibits mitorespiration with no effect on chlororespiration. Estimation of the chlororespiration rate, from oxygen consumption measurements with these different inhibitors, gives values between 35 and 70 nmoles O_2 min^{-1} chl, i.e. 10 to 20 % of total respiratory activity.

The photoinhibition of the chlororespiration activity which was observed under flash excitation is mediated by photosystem I, as shown by the

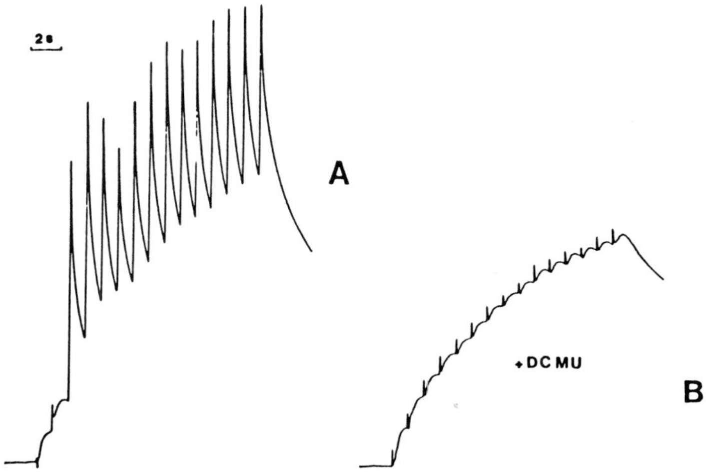

Fig. 5: Amperometric signal measured following a flash excitation of Chlamydomonas cells deposited on a bare platinum electrode. A: cells in the presence of 1 mM acetate. B: after adition of 10 uM DCMU to inhibit O_2 evolution.

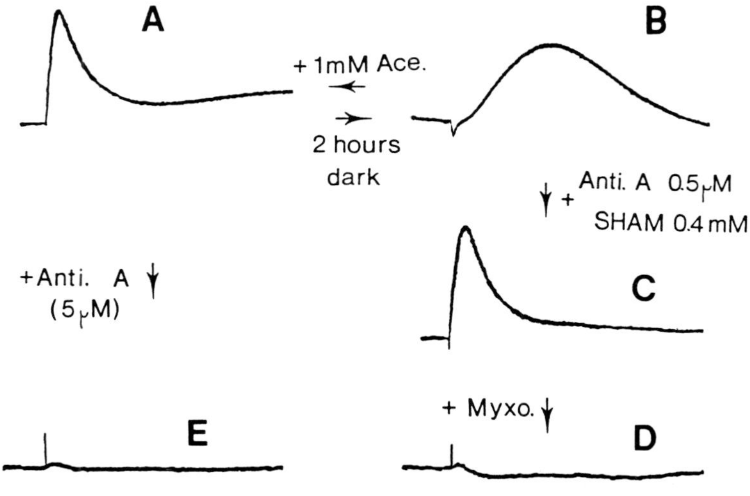

Fig. 6: Amperometric signals in Chlamydomonas following a single actinic flash in presence of 1 mM acetate (A), after 2 hours in the dark (B), after addition of 0.5 uM antimycin A plus 0.4 mM SHAM (C) and after subsequent addition of 5 uM myxothiazol (D) or 5 uM antimycin (E).

insensitivity of this phenomenon to DCMU (Fig.4B) and its absence in a mutant of Chlamydomonas (F15/4+) lacking photosystem I. This light-induced inhibition of chlororespiration has been interpreted as being due to the oxidation, by photosystem I activity, of an electron carrier common to both the chloroplastic and chlororespiratory chain (Peltier et al., 1987). Using different mutants of Chlamydomonas Bennoun (1983) has shown that electron transfer components located between the PQ pool and photosystem I reaction center (cyt b, cyt f, plastocyanin) were not involved in the oxidation of plastoquinone by the chlororespiratory pathway. The results suggest that changes in the redox state of the PQ pool, induced by photosystem I, are responsible for the light inhibition of chlororespiration. No correlation is observed between the rates of the light-induced inhibition of the chlororespiration and those of the electron transfer reactions to photosystem I. For example, after consumption of the endogenous pool of substrates the half-time for the light-induced inhibition of chlororespiration is about 2000 ms (Fig. 6) (Peltier and Ravenel to be published). If acetate is added, or if the PQ pool is partly reduced by addition of mitochondrial inhibitors, a half-time of 300 ms is observed (Fig. 6). In both cases, the half-time of light-inhibition of chlororespiratory activity is much lower than that of electron transfer between cytochrome f and P700 (100 to 600 µs, Delosme R., personal communication).

Peltier and Ravenel (unpublished results) have observed that after addition of high concentrations (5 µM) of antimycin A or myxothiazol, no inhibition of chlororespiration can be observed under a single actinic flash (Fig.6). However, such inhibitory concentrations had no effect on the electron transfer in the cyt b6f complex as inferred by the fact that the slow phase of the carotenoid shift was unchanged (not shown). Two hypotheses can be put forward to explain this observation and other available data:

1. The terminal oxidase is directly branched on the PQ pool. This oxidase is inhibited by antimycin A and myxothiazol at high concentrations.

2. There are two types of cytochrome bc complexes: the b6f complex involved in the photosynthetic pathway - not sensitive to antimycin A and myxothiazol - and a bc complex linked to chlororespiration which is sensitive to these chemicals.

In any case the interaction between the chlororespiration and photosynthetic electron transfer chain appears to occur at the level of the PQ pool in a similar manner to the one proposed for photosynthetic bacteria (Lavorel et al., 1989). This is summarized in Fig. 7.

The chemical nature of the chloroplast oxidase(s) is still unknown. Rolfe et al. (1987) have proposed that the new cytochrome b which they have discovered in both *Chlamydomonas* and higher plants chloroplasts is involved in the chlororespiration pathway. The mid-point redox potential value of -125 mV measured for that cytochrome is however very low for an oxidase Two heme proteins, H_1 and H_2, found by Lemaire et al. (1986) also appear as possible candidates. In this context, Peltier and Schmidt (to be published) have observed a drastic stimulation of chlororespiratory activity in *Chlamydomonas* cells growing under nitrogen limitation. This correlates with an increase in the thylakoid concentration of the protein H_1.

III. CHLORORESPIRATION AND THE KOK EFFECT

The lack of linearity in the curve of net oxygen exchange as a function of light-intensity, or "Kok effect", has been interpreted as related to the inhibition of dark respiration mediated by the photochemical reactions (Kok, 1949). Because of the insensitivity of the "Kok effect" to DCMU, and its wavelength dependence, it was proposed that the inhibition of respira-

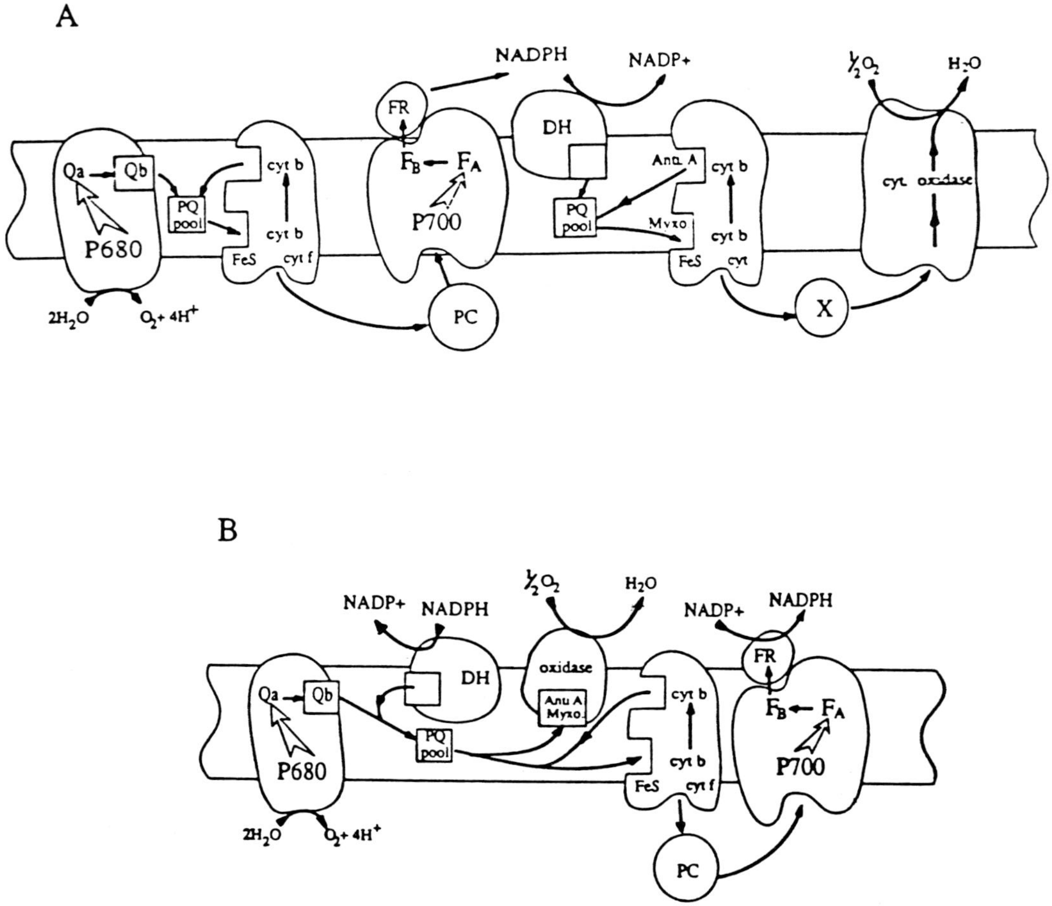

Fig. 7: Schemes of the possible interactions between chloroplastic and photosynthetic electron transfer chains in Chlamydomonas.

tion was induced by the ATP generated by cyclic photophosphorylation around photosystem I (Hoch et al., 1963). This hypothesis is, however, inconsistent with the insensitivity of the "Kok effect" to uncouplers like CCCP. Healey and Myers (1971) therefore proposed that oxygen uptake was depressed by a diversion of reductants away from repiratory electron transport and into photosystem I. Peltier and Sarrey (1988), measuring oxygen exchange by mass spectrometry, have shown that, in conditions where the mitochondrial activity was inhibited, the "Kok effect" was still present. They concluded therefore that the "Kok effect" is not a consequence of the inhibition of mitochondrial activity but is due to the inhibition of the chlororespiration induced by photosystem I activity.

IV. EVIDENCE FOR CHLORORESPIRATION IN HIGHER PLANTS

Although a clear-cut demonstration is still lacking, several lines of evidence indicate the occurrence of chlororespiration in higher plants. First of all, the "Kok effect" has been detected in many higher plants (Kok, 1949; Ishii et al., 1977; Sharp et al., 1984). As previously discussed, this effect is at least partially related to the inhibition of chlororespiration activity induced by photosystem I photochemistry.

The presence in the chloroplast genome of genes homologous to the components of NADH dehydrogenase, that are encoded in mitochondrial DNA, has been reported for tobacco (Shirozaki et al., 1986) and a liverwort (Okyama et al., 1986). These chloroplastic NADH dehydrogenase genes are not inactive copies of mitochondrial genes that have migrated between different cell compartments during evolution, since they are actively expressed in tobacco chloroplasts (Matsubayashi et al., 1987). More recently a gene encoding a homologe of a nuclear coded component of the mitochondrial

NADH dehydrogenase has also been characterized in the chloroplastic DNA (Fearnley et al., 1989).

By fluorescence yield measurements and an analysis of electrochromic absorbance changes, Garab et al. (1989) have recently shown that the addition of KCN (0.5 to 5 mM) induces a large reduction of the PQ pool in tobacco leaves and protoplasts, as well as in intact isolated pea chloroplasts. These results suggest strongly that NAD(P)H-plastoquinone-oxidoreductase is indeed functioning in higher plant chloroplast thylakoids.

V. WHAT IS THE PHYSIOLOGICAL ROLE OF CHLORORESPIRATION ?

Since it is usually supposed that green plant cells have evolved from a symbiosis between an eukaryotic host cell and a photosynthetic prokaryote, chlororespiration could be regarded as the residual respiratory activity of the latter one. On the other hand, as proposed by Bennoun (1982), chlororespiration can ensure the recycling of NAD(P)H formed during starch degradation by the glycolytic pathway. The first step in this process, i.e. transformation of 3,glucose-6-phosphate into 3,gluconate-6-phosphate, is tightly controled by NAD(P)H. The reoxidation of the NAD(P)H formed during glycolysis by the chlororespiration could be important for the regulation of this process in the dark.

Another possible function of chlororespiration is to sustain a sufficient electrochemical gradient in order to maintain the ATPase in a functional form (Shahak, 1985), even after a long dark period.

Nie et al. (1987) have found that an iron-sulphur protein, a subunit of the chloroplast NADH dehydrogenase, could form a stable DNA protein complex with a cloned chloroplast DNA of replicative origin. They proposed a possible implication of chlororespiration in regulating the initiation of chloroplast DNA replication (Wu et al., 1989).

VI. CONCLUSIONS

If the entry of electrons into the thylakoid membranes of eukaryotic photosynthetic organisms appears to be well documented, biochemical proper ties of the enzyme(s) associated with that portion corresponding to the chlororespiration electron transfer are scare and not well established. The situation is worst with respect to the oxidase(s) responsible for the oxidation of the PQ pool. Neither the proteins involved nor their related genes have been characterized. Further research concerning chlororespiration will have to provide a clear-cut demonstration of electron transfer from substrates to molecular oxygen in intact chloroplasts isolated from both micro-algae and higher plants. A more detailed biochemical characterization of the nature, the determination of the number of electron transfer components (dehydrogenase, complex bc, oxidase etc.) involved in chlororespiratory activity are also necessary for precisely defining the function and eventual regulation of this process.

REFERENCES

Ben-Amotz A and Gibbs M, (1975) Hydrogen metabolism in photosynthetic organisms. II Light-dependent hydrogen evolution by preparation from *Chlamydomonas*, *Scenedesmus* and *Spinach*. Biochim. Biophys. Res. Commun. 64:355-359.

Bennoun P (1982) Evidence for a respiratory chain in the chloroplast Proc. Natl. Acad. Sci. USA 79:4352-4256.

Bennoun P (1983) Effect of mutations and ionophores on chlororespiration in *Chlamydomonas reinhardtii*, FEBS Lett. 156:363-365.

Binder A, Hauser R and Krogmann D (1984) Respiration in energy transducing membranes in a thermophilic cyanobacterium. In: Selman BR and Selman-Reimer S (eds) Energy coupling in photosynthesis. Elsevier/ North Holland,New-York, 313-321.

Böhme H, Reimer S and Trebst A (1971) The role of dibromothymo-quinone, an antagonist of plastoquinone on non cyclic and cyclic electron flow in isolated chloroplast. Naturforsch.26b:341-352.

Cotton NPJ, Clark AJ and Jackson JB (1983) Interaction between the respiratory and photosynthetic electron transport chains of intact cells of *Rhodopseudomonas capsulata* mediated by membrane potential. Eur.J. Biochem. 130:581-587.

Fearnley IM, Runswick MJ and Waller JE (1989) A homologous of the nuclear coded 49Kd subunit of bovine mitochondrial NADH ubiquinone reductase is coded in chloroplast DNA. EMBO.J. 8:665-672.

Gaffron H and Rubin J (1942) Fermentative and photochemical production of hydrogen in green algae. Am. J. Bot. 27:273-283. Gfeller, R.P. and Gibbs M. (1985). Fermentative metabolism of *Chlamydomonas reinhardtii* .II role of plastoquinone 77:509-511.

Garab G, Lajko F, Mustardy L and Marton L (1989) Respiratory control over photosynthetic electron transport in chloroplasts of higher plants cells: evidence for chlororespiration. PLanta, in press.

Gfeller RP and Gibbs M (1984) Fermentation metabolism of *Chlamydomonas reinhardtii*.I Analysis of fermentative products from starch in dark and light.Plant Physiol. 75:212-218.

Gibbs M, Gfeller RP and Chen C (1986) Fermentative metabolism of *Chlamydomonas reinhardtii*. III Photoassimilation of acetate. Plant. Physiol. 82:160-166.

Godde D and Trebst A (1980) NADH as electron donor for the photosystem membrane of *Chlamydomonas reinhardtii*. Arch.Microbio.127:245-252.

Goedheer JC, (1963) A cooperation of two pigment systems and respiration in photosynthetic luminescence. Biochim. Biophys. Acta. 66:61-71.

Healey FP (1970)a. Hydrogen evolution by several algae. Planta 91:220-226.

Healey FP (1970)b. The mechanism of hydrogen evolution by *Chlamydomonas reinhardtii*. Plant Physio. 45:153-159.

Healey FP and Myers J (1971) The Kok effect in *Chlamydomonas reinhardtii*. Plant Physiol. 47:373-379.

Hirano M, Satoh K and Katoh S (1980) Plastoquinone as a common link between photosynthesis and respiration in a blue-green alga. Photosynth. Res. 1:149-162.

Hoch GO, Owens H and Kok B (1963) Photosynthesis and respiration. Arch. Biochim. Biophys. 101:171-180.

Ishii R, Takehara T, Murata Y and Miyachi S (1977) Effect of light intensity on the rates of photosynthesis and photorespiration in C-3 and C-4 plants Biological Solar Energy Conversion, Academic Press, New-York, p 265.

Kaltwasser H, Stuart TS and Gaffron H (1969) Light dependent hydrogen evolution by *Scenedesmus*. Planta 89:309-321.

Knaff DB (1978) Reducing potentials and the pathway of NAD^+ reduction in "The photosynthetic bacteria" edited by RK Clayton and WR Sistrom. Plenum Press New York and London p.629-638.

Kok B (1949) On the interrelation of respiration and photosynthesis in green plants. Biochim. Biophys. Acta 3:625-631.

Lavorel J, Richaud P and Vermeglio A (1989) Interaction of photosynthesis and respiration in *Rhodospirillaceae* evidence for two functionally distinct bc_1 complex fractions. 973:290-295.

Lemaire C, Girard-Bascou J, Wollman FA and Bennoun P Studies on the cytochrome b_6/f complex. Characterisation of the complex subunits in *Chlamydomonas reinhardtii*. Biochim.Biophys.Acta 851:229-238.

Lockau W (1981) Evidence for a dual role of cytochrome C553 and plastocyanin in photosynthesis and respiration of the cyanobacterium, *Anabaena variabilis*. Arch. Microbiol. 128:336-340.

McCarthy JEG and Fergusson SJ (1982) Respiratory control and the basis of light-induced inhibition of respiration in chromatophores from *Rhodopseudomonas capsulata*. Biochim.Biophys. Res. Commun. 107:1406-1411.

Matsubayaski T, Wakasugi T, Shinozaki K, Yamaguchi-Shirozaki K, Zaita N, Hidata T, Meng BY, Ohta C, Tanaka M, Kato A, Maruyama T, and Sugiura M

(1987) Six chloroplast genes (ndh A-F) homologous to human mitochondrial genes encoding components of the respiratory chain NADH deshydrogenase are actively expressed: determination of the splice sites in ndh A and ndh B pre-m RNAL. Mol. Gen. Genet. 210:385-393.

Matthijs HCP, Luderus EME, Löffler HJM, Scholts MJC and Kraayenhof R (1984) Energy metabolism in the cyanobacteria *Plectonema boryanum*. Participation of the photosynthetic electron transfer chain in the dark respiration of NADPH and NADH.Biochim.Biophys. Acta 766:29-37.

Nakamura H (1937) Über die photosynthese beider schwefelfreien purpurbakterie *Rhodobazillus palustris*. Beiträge zur stoffwechselphysiologie der purpurbakterie. I, Acta Phytochim.,9:189-234.

Nie ZQ, Chang DY and Wu M (1987) Protein-DNA interaction within one cloned chloroplast DNA replication origin of *Chlamydomonas*. Mol. Gen.Genet., 209:265-269.

Peltier G, Ravenel J and Vermeglio A (1987) Inhibition of a respiratory activity by short saturating flashes in *Chlamydomonas* : evidence for a chlororespiration. Biochim.Biophys.Acta 893:83-90.

Peltier G and Sarrey F (1988) The Kok effect and the light inhibition of chlororespiration in *Chlamydomonas reinhardtii*.FEBS Lett. 228:259-262.

Peltier G and Thibault P (1988) Oxygen-exchange studies in *Chlamydomonas* mutants deficient in photosynthetic electron transport: evidence for a Photosystem II-dependent oxygen uptake in vivo. Biochim.Biophys. Acta 936:319-324.

Rebeille F and Gans P (1988) Interaction between chloroplast and mitochondria in microalgae. Role of glycolysis. Plant Physiology 88:973-975.

Richaud P, Marrs BL and Vermeglio A (1986) Two modes of interactions between photosynthetic and respiratory chains in whole cells of *Rhodopseudomonas capsulata*. Biochim.Biophys.Acta. 850:256-263.

Rolfe SA, Sanguansermsri M and Bendall DS (1987) Cytochrome b560, a new component of thylakoïd membrane. Biochim.Biophys.Acta 894:434-442.

Sahak Y, (1985) Different effect of thioloxidants on the chloroplast H^+-ATPase in the light and in the dark. J. Biol. Chim. 260:1459-1464.

Scherer S, Almon H and Böger P (1988) Interaction of photosynthesis respiration and nitrogen fixation in cyanobacteria. Photosynth.Res. 15:95-114.

Sharp RE, Mathews MA and Boyer JS (1984) Kok effect and the quantum yield of photosynthesis. Plant Physiol. 75:95-101.

Stuart TS and Gaffron H (1971) The kinetics of hydrogen photoproduction by adapted *Scenedesmus*. Planta 100:228-243.

Van Niel CB (1941) The bacterial photosynthesis and their importance for the general problem of photosynthesis. Adv. Enzymol. 1:263-328.

Willeford KO, Gombas Z and Gibbs M (1989) Evidence for chloroplastic succinate deshydrogenase participating in the chloroplastic respiratory and photosynthetic electron transport chains of *Chlamydomonas reinhardtii*. Plant. Physiol. 90:1084-1087.

Wu M, Nie ZQ and Yang Y (1989) The 18kD protein that binds to the chloroplast DNA replicative origin is an iron-sulfur protein related to a subunit of NADH deshydrogenase. The Plant Cell,1:551-557.

Electron transfer in the oxygen-evolving system II of photosynthesis

E. Schlodder
Max-Volmer-Institut für Biophysikalische
und Physikalische Chemie
Technische Universität Berlin
Str. d. 17. Juni 135
1000 Berlin 12
FRG

Introduction

In photosystem II (PS II) the sequence of light-driven electron transfer reactions results utlimately in the oxidation of water to O_2 and the reduction of plastoquinone. Recent reviews on photosystem II and oxygen evolution include, among others, articles by Babcock (1987), Mathis and Rutherford (1987), Witt et al. (1986) and Hansson and Wydrzynski (1989).

After absorption of light by an antenna pigment, the excitation energy is channeled to a specialized chlorophyll a (P680 or Chl-a_{II}), the primary electron donor of the reaction center of PS II (Döring et al., 1967). From the singlet excited state of P680, the primary charge separation takes place leading to the oxidation of P680 and the reduction of a pheophytin-a molecule (for review see Klimov and Krasnovsky, 1981). The charge separation is stabilized by secondary electron transfer from the reduced pheophytin to a special plastoquinone, Q_A (Stiehl and Witt, 1968). The formation of $P680^+Q_A^-$ takes less than one nanosecond (Nuijs et al., 1986; Schatz et al., 1988; Gerken et al, 1987). Subsequently, the semiquinone anion, Q_A^-, transfers its electron in a few hundred microseconds to another plastoquinone, Q_B, which functions as a two-electron gate. Q_B^- remains tightly bound to the reaction center until the next light-induced charge separation leads to the reduction of Q_B^- to $Q_B^=$. The doubly reduced quinone takes up two protons and leaves as plastoquinol, QH_2, the reaction center to be replaced by another neutral plastoquinone (for review see Crofts and Wraight, 1983).

On the donor side the very strong oxidant, $P680^+$ ($E_m \approx +1.14$ eV (Klimov and Krasnovsky, 1981)), which is capable of oxidizing water to O_2, is re-reduced by a secondary donor, Z, probably tyrosine-161 of the D1-polypeptide, one of the reaction center proteins (see Results and Debus et al., 1989 and Gerken et al., 1988). The oxidized Z is, in turn, reduced by the oxygen-evolving complex, S (Babcock et al., 1976; Dekker et al., 1984) that contains a cluster of four manganese ions. S accumulates four oxidizing equivalents which are produced by four consecutive photo-oxidations of P680. It thereby passes through four different redox states (the so-called S states $S_o \rightarrow S_3$) before O_2 is released during the transition $S_3 \rightarrow S_o$ (Joliot et al., 1969; Kok et al.,1970). In this way

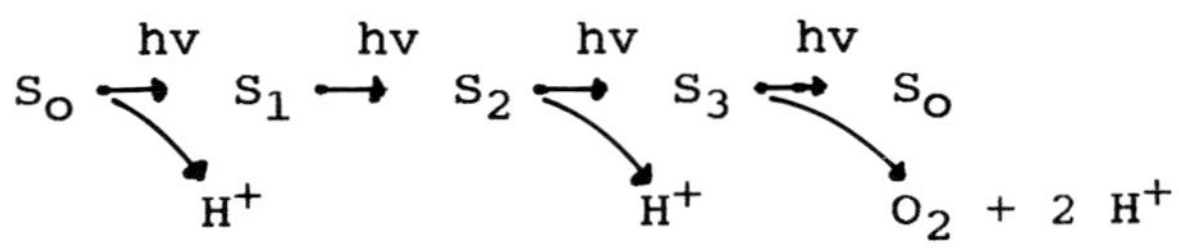

the one-electron photo-oxidation of P680 is matched to the overall four-electron process of water oxidation. The oscillation is damped due to losses in the accumulation of oxidizing equivalents ("misses") and double turnovers ("double hits"). In dark-adapted samples the O_2-evolving complexes are in the state S_1 (Vermaas et al., 1984). The successiv redox transitions $S_o \rightarrow S_1 \rightarrow S_2 \rightarrow S_3 \rightarrow S_o$ are accompanied by proton release. The sequence of proton release measured around pH 7 (the optimum of O_2 evolution) is 1, 0, 1, 2 (Fowler, 1977; Förster and Junge, 1985).

During the past ten years progress in PS II research has been closely related with the development of techniques for the separation of PS II from other protein complexes in the photosynthetic membrane. Isolated and purified O_2-evolving PS II complexes from blue green algae are particularly stable and well suited for the functional analysis of PS II.

The polypeptide composition of PS II complexes from cyanobacteria is very similar to that of higher plants (Ghanotakis and Yocum, 1985; Satoh, 1985). The reaction center that binds the components of the primary electron transfer chain consists of the D1 (32 kDa), D2 (34 kDa) and the cyt-b-559 (9 and 4 kDa) proteins (Namba and Satoh (1987)). Chl-a antenna pigments are bound to the polypeptides CP 47 and CP 43. Additionally, low molecular mass polypeptides have been detected. In higher plants, three extrinsic polypeptides (33 kDa, 23 kDa and 17 kDa) are present and involved in the process of water oxidation. The 17 and 23 kDa proteins are absent in cyanobacteria (Stewart et al., 1985). These proteins play a role in the requirement of Cl^- and Ca^{2+} for oxygen evolution in higher plants (for review see Homann, 1987).

In this work the analysis of the electron transfer between the oxidized primary electron donor, P680+, and the oxygen-evolving complex by time-resolved absorbance difference spectroscopy is described in more detail. Isolated PS II complexes from the thermophilic blue green alga Synechococcus sp. have been used for these studies.

Materials and Methods

Cells of the thermophilic cyanobacterium Synechococcus sp. were grown in two 30 l-glass columns at 57 °C. Oxygen-evolving PS II complexes were extracted in principle as described by Schatz and Witt (1984). Membrane fragments, prepared from the cells, were treated with the zwitterionic detergent sulfobetaine 12. After centrifugation of the membrane suspension at 140,000 x g for 40 min, the supernatant contained the solubilized PS II complexes. For some experiments the extract was further purified by sucrose density-gradient centrifugation (Dekker et al., 1988), in order to remove most of the phycobiline pigments. The PS II preparations were stored at -80 °C for further use.

These preparations offer advantages for spectroscopic studies: (a) The signal-to-noise ratio is substantially improved due to the smaller antenna size (about 60 Chl-a per PS II based on the amount of photo-reducible Q_A). (b) There is practically no interference with signals due to PS I because of the low content of PS I (PS II : PS I $\geq$ 20). (c) Light scattering is strongly reduced compared to membrane fragments. (d) The PS II complexes are more stable compared to similar preparations from higher plants. At 4 °C the oxygen-evolving activity decreases only by approx. 10% within 24 hours.

The steady state O_2-flash yield was about $(2.5 - 3.6) \cdot 10^{-3}$ O_2 per flash and Chl, corresponding to (70-100) Chl per PS II active in O_2 evolution. This indicates that a fraction of the PS II complexes (15-40%) have lost their O_2-evolving capacity in the course of the preparation.

Absorption changes, following excitation of the sample by short laser flashes, lead to changes in the transmitted measuring-light intensity. The kinetics and the spectrum of these changes were measured and attributed to the light-induced electron-transfer reactions. For different time ranges and wavelength regions between 250 nm and 1000 nm purpose-built spectrophotometers were used as described elsewhere (Schlodder and Brettel, 1988). Samples were excited at 532 nm by flashes of 3 ns- or 35 ps duration from a fequency-doubled Nd/YAG laser. The fastest time resolution obtained by the detection system was about 1 ns. In order to improve the signal-to-noise ratio, the signals were induced repetitively and were averaged.

Flash-induced oxygen evolution was measured with a zirconia oxygen sensor as described in (Schlodder and Meyer, 1987).

Results and Discussion

$P680^{\pm}$ reduction kinetics

Oxidation and re-reduction of the primary donor, P680 or Chl-a_{II}, can be followed around 680 and 824 nm (van Best and Mathis, 1978; Schlodder et al., 1984). Fig. 1 shows the time course of the absorption change at 824 nm under repetitive excitation of PS II complexes from Synechococcus. The flash-induced absorption increase reflects the formation of $P680^{+}$ that has a broad absorption band in the near-infrared, similar to the monomeric Chl-a cation radical. The re-reduction kinetics of $P680^+$ are clearly very complex (see Fig. 1). Analyses of time course in the ns to 100 ms

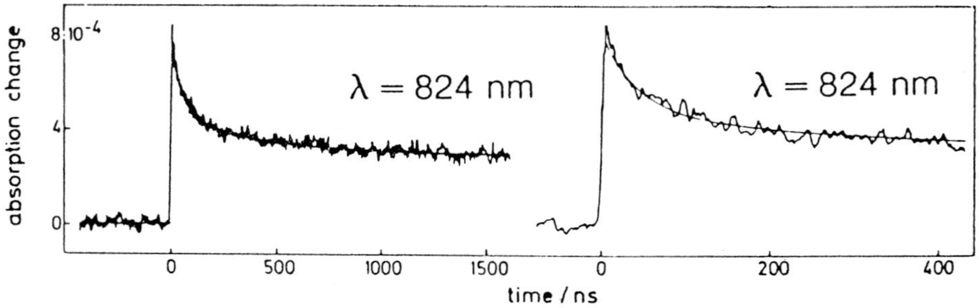

Fig. 1: Time course of flash-induced absorption changes in the nanosecond time range attributed to P680 photo-oxidation and re-reduction in O_2-evolving PS II complexes from Synechococcus sp.; pH 6.5, $2 \cdot 10^{-4}$ M phenyl-p-benzoquinone as electron acceptor, electrical bandwidth: 100 Hz-100 MHz. The solid line was calculated using three exponential phases with $t_{1/2}$ = 20 ns (30%), 50 ns (19%) and 280 ns (19%) and slower reduction phases (32%).

time range (not shown) by a sum of exponentials yielded phases with half-life times between 20 ns and 5 ms. The complex kinetics indicate that the state of the donor side (e.g., the presence of an intact or damaged O_2-evolving complex, the S state of the O_2-evolving complex) and conditions of reaction medium (e.g., pH) strongly influence the rate of $P680^+$ re-reduction.

In intact O_2-evolving complexes the electron donation from the secondary donor, Z, to $P680^+$ occurs with kinetic phases in the nanosecond time range and with a minor phase of about 35 µs (Brettel et al., 1984; Schlodder, et al., 1985). Three ns phases with half-life times of approx. 20 ns, 50 ns and 260 ns could be distinguished measuring 824 nm abosorption changes in single flashes. Fig. 2 shows the time course of $P680^+$ re-reduction, observed after each flash of a series of four saturating laser flashes, given to a dark-adapted samples of PS II complexes from Synechococcus (pH = 6.5). After the first and fifth (not shown) flash, when the S states S_o or S_1 are present, $P680^+$ reduction is nearly monophasic with

$t_{1/2} \approx 20$ ns. After the second and third flash, when S_2 or S_3 is dominantly present, slower biphasic kinetics ($t_{1/2}$ = 50 ns and 260 ns) were found. After the fourth flash, when S_2/S_3 as well as S_o/S_1 are present prior to the flash, the kinetics are intermediate (see Fig. 2). The multiphasic nanosecond reduction under repetitive excitation (see Fig. 1) can be explained quantitatively by a superposition of the different kinetics related to the four S states.

The retardation of electron transfer in the higher S states, S_2 and S_3, can be explained by an electrostatic effect of a positive excess charge located in the O_2-evolving complex (Brettel et al., 1984). A positive net charge in S_2 and S_3 results, if the electron and proton release patterns are 1,1,1,1 and 1,0,1,2, respectively, for the sequence of S-state transitions ($S_o \rightarrow S_1 \rightarrow S_2 \rightarrow S_3 \rightarrow S_o$).

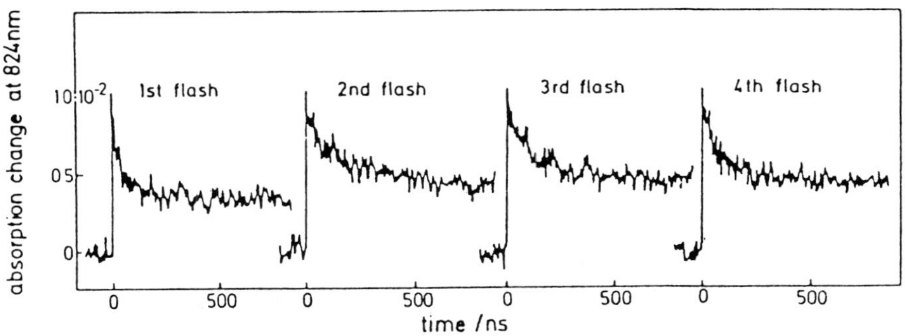

Fig. 2: Time course of the absorption change at 824 nm after each flash of a series of four saturating laser flashes given to dark-adapted samples of PS II complexes from Synechococcus

This explanation is supported by studies of the P680$^+$ nanosecond reduction kinetics in single flashes as a function of pH (Meyer et al., 1989). Between pH 8.0 and 6.0 the kinetics are nearly pH-independent. Below pH 6.0 the half-life time after the first flash increases from about 20 ns to about 45 ns at pH 4.0, and the biphasic reduction, after the second flash with $t_{1/2}$ = 50 and 260 ns, changes to nearly 100% with $t_{1/2} \approx 260$ ns. It seems reasonable to assume that the slower kinetics upon lowering of the pH reflect an increase of positive net charge due to a protonation at acidic pH, which would give rise to a higher Coulomb attraction. If the number of positive excess charges depends on the oxidation- and protonation state of the oxygen-evolving complex, the biphasic nanosecond re-reduction kinetics of $P680^+$ in states S_2 and S_3 (see above) might be explained by the coexistence of two different protonation states of S.

After reversible (e.g., by low pH-, or acetate treatment) or irreversible (e.g., by Tris- or heat treatment) inhibition of oxygen evolution, $P680^+$ is reduced by the same donor (see below), but the kinetics are slowed down by a factor of about 1000 from the ns- to the µs time range. The reduction kinetics depend on pH ($t_{1/2} \approx 4$ µs at pH 6.5, 30 µs at pH 4.5 and 0.2 µs at pH 9.0)(Conjeaud and Mathis, 1986; Schlodder and Meyer, 1987). It is poorly understood which kind of modification alters the kinetic properties of Z. Among others a shift of the Z^+/Z redox potential following inhibition (Yerkes et al., 1983) or conformational changes may be the reason.

The slowest phases in $P680^+$ re-reduction with half-life times of 170 µs, 800 µs and about 6 ms occur, if a charge separation between P680 and Q_A is induced by the flash, and Z is still oxidized. Under these conditions, $P680^+$ is re-reduced by the charge recombination with Q_A^- (Gerken et al., 1989). The reason for the multiphasic charge recombination is not quite clear. Different structural states of the reaction-center protein and/or a heterogeneity of the acceptor side (Q_B or Q_B^- present) might be proposed as explanation.

After the different reduction phases of $P680^+$ have been assigned, the kinetics can be used as a probe for alterations of the donor side induced by specific treatments. For example, the ion requirement for O_2 evolution has been analyzed in this way. For higher plants, evidence has been presented that chloride and calcium are specifically required in the process of water oxidation (Homann, 1987). Significant differences compared to higher plants have been observed in PS II complexes from Synechococcus (Meyer, 1987). If Cl^- and Ca^{2+} are removed by gel filtration (the reaction medium contains only 50 mM MES-buffer (pH 6.5) and 20 mM Na_2SO_4), the ns reduction phases that reflect the O_2-evolving activity disappear almost completely (see Fig. 3, trace a). The activity (ns-reduction phases) can be restored by addition of $CaCl_2$ (Fig. 3, trace d). So far, the results are in agreement with those reported for higher plants. However, other salts, containing either no Ca^{2+} (e.g., 250 mM NaCl (Fig. 3, trace c) or no Cl^-, restore the activity. Even after addition of 20 mM $MgSO_4$ (see Fig. 3, trace b) or high amounts of Na_2SO_4 (1 M)(not shown) a re-activation is observed (Pauly et al., 1989). At least for PS II from cyanobacteria these results make a specific function of calcium and chloride in the mechanism of water oxidation rather unlikely.

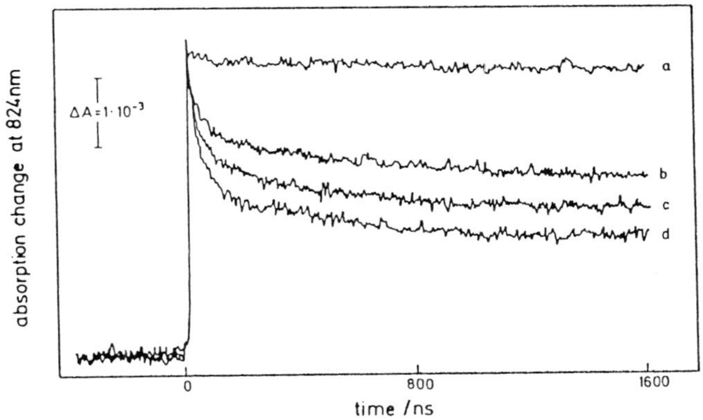

Fig. 3: Time course of flash-induced absorptions changes at 824 nm: (a) after removal of Cl^- and Ca^{2+}. The reaction medium contains $50 \cdot 10^{-3}$ M Mes-NaOH (pH 6.5) and $20 \cdot 10^{-3}$ M Na_2SO_4, and after addition of (b) $20 \cdot 10^{-3}$ $MgSO_4$, (c) $250 \cdot 10^{-3}$ NaCl and (d) $20 \cdot 10^{-3}$ $CaCl_2$.

<u>Characterization of the immediate donor to $P680^{\pm}$</u>

Absorption changes due to the electron transfer reaction $P680^+ \; Z \rightarrow P680 \; Z^+$ contain (a) the contribution due to the rereduction of $P680^+$ and (b) the contribution due to the oxidation of Z, which is negligible in the red and near-infrared wavelength region. In the UV both contributions are superimposed. Monitoring absorption changes in the UV with nanosecond time resolution therefore offers the possibility to characterize the immediate donor to $P680^+$ in O_2-evolving preparations (Gerken et al., 1988). At all wavelengths the nanosecond decay could be adapted with the same multiphasic exponential kinetics as those observed at 824 nm (see Fig. 1), where absorption changes are exclusively due to $P680^+$ reduction. The oxidized immediate donor is, in turn, re-reduced with a half-life time greather than 10 µs. Fig. 4A depicts the amplitude of the ns phases as a function of wavelength between 250 nm and 330 nm. After inhibition of oxygen evolution by Tris-treatment, the electron donation from Z to P680+ occurs with a half-life time of several microseconds (13 µs at pH 5.2). Fig. 4B shows the spectrum of the 13 µs phase. Both spectra are practically identical. From this result one can conclude that in O_2-evolving PS II complexes and in Tris-treated PS II complexes (O_2 evolution inhibited) the immediate donors to $P680^+$ are of the same chemical nature. The altered kinetics are most probably due to a modification of Z or structural changes induced by inhibition of O_2 evolution.

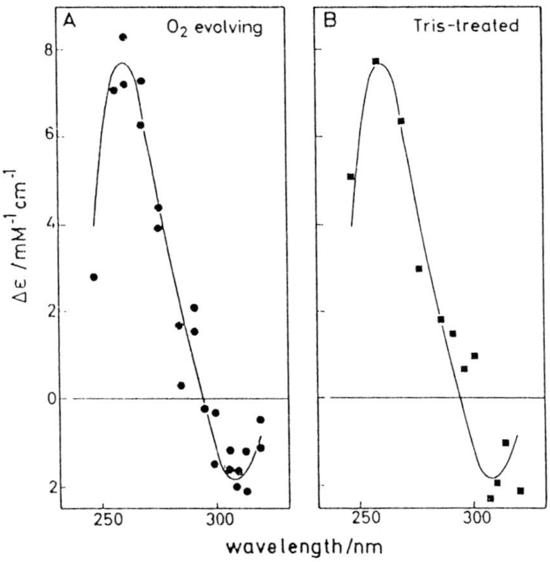

Fig. 4: Difference spectrum for the electron transfer reaction $P680^+\ Z \rightarrow P680\ Z^+$ in O_2-evolving PS II complexes (A) and in PS II complexes with inhibited O_2 evolution (B).

In order to evaluate the difference spectrum attributed to the oxidation of only the immediate donor, the contribution due to P680+ → P680 has been subtracted from the total spectra shown in Fig. 4. The result is given in Fig. 5A or 5B, respectively. For a comparison in Fig. 4C, the in vitro difference spectrum of the oxidation of tyrosine in water is shown (taken from Bent and Hayon, 1975). The similarity of the spectra supports the identification of Z as a tyrosine (Gerken et al., 1988). This has been suggested at the same time on the basis of EPR experiments using specific deuteration (Barry and Babcock, 1987) and studies using site-directed mutageneses (Debus et al., 1988). More recently, evidence has been presented that it is tyrosine-161 in the D1 polypeptide functioning as intermediate electron carrier between $P680^+$ and the oxygen-evolving complex (Debus et al., 1989).

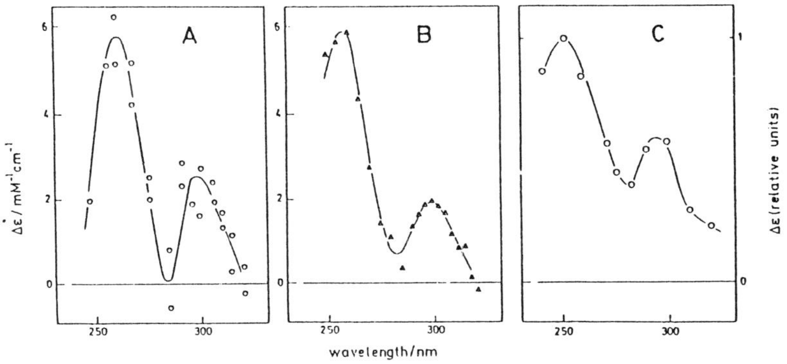

Fig. 5:
A,B: Difference spectrum of the oxidized minus reduced form of the immediate donor to $P680^+$ in O_2-evolving PS II complexes (A) and in Tris-treated PS II complexes (B).
C: In vitro difference spectrum taken from Bent and Haydon (1975) attributed to the reaction tyrOH → tyrO$^\bullet$ + H^+

Assuming similar pK values for Z as for tyrosine in water, the oxidation of Z should be accompanied by a deprotonation (pK < 0 for tyrO$^\bullet$H$^+$ ⇌ tyrO + H^+; OH designates the phenolic hydroxyl group) and, in turn, the re-reduction should be coupled with a protonation (pK ≈ 10 for tyrOH ⇌ tyrO$^-$ + H^+). H^+-release and re-uptake concomitant with oxidation and reduction of the donor has been reported for inactivated PS II complexes (Renger and Völker, 1982), but have yet not been observed for O_2-evolving PS II.

The pH-dependent redox potential for tyrosine in water is E_m (tyrO$^\bullet$/tyrOH) = 1340 mV - 59 mV·pH for pH < 10 (Harriman, 1987). This yields a redox potential of 930 mV at pH 7.0; i.e., about 200 mV more negative than that of P680/P680$^+$ (E_m ≈ 1.14 V (Klimov and Krasnovksy, 1981)) and positive enough to drive the oxidation of water (E_m (O_2, 4 H^+/2 H_2O) ≈ 1229 mV - 59 mV·pH,i.e.; 816 mV at pH 7.0). In Fig. 6 the redox potentials are depicted as funciton of pH. The redox potential of the $S_1 \rightarrow S_2$ transition should be pH-independent, because no H+ is released during this transition, and it has been estimated to be 850-900 mV (Bouges-Bocquet, 1980).

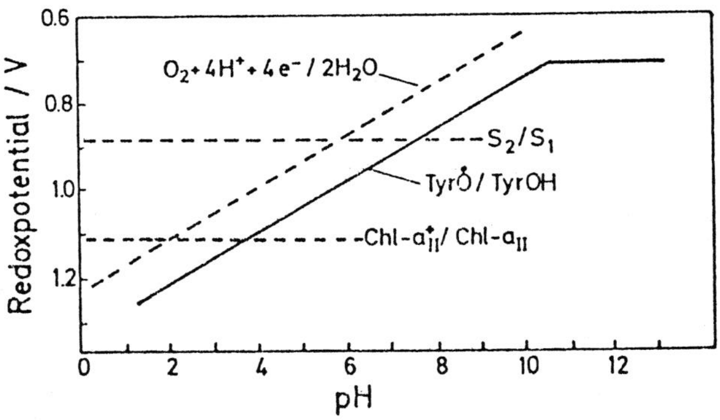

Fig. 6: Redox potentials of the indicated half reaction vs. normal hydrogen electrode (see text for details and references)

The redox potential scheme shown in Fig. 6 offers an explanation for the observed pH dependence of O_2 evolution (Schlodder and Meyer, 1987). Fig. 7 shows the dependence of the O_2-flash yield on the pH measured with a zirconia oxygen sensor for PS II complexes from Synechococcus. At low pH O_2 evolution shows a half maximum at about 4.5. The reason for the reversible inhibition upon lowering of the pH might be that P680$^+$ can no longer oxidize Z (see Fig. 6). Under these conditions, P680$^+$ would be reduced by a charge recombination with Q_A^-. An increase of P680$^+$-decay phases with half-life times ≥ 150 µs is indeed observed below pH 5.

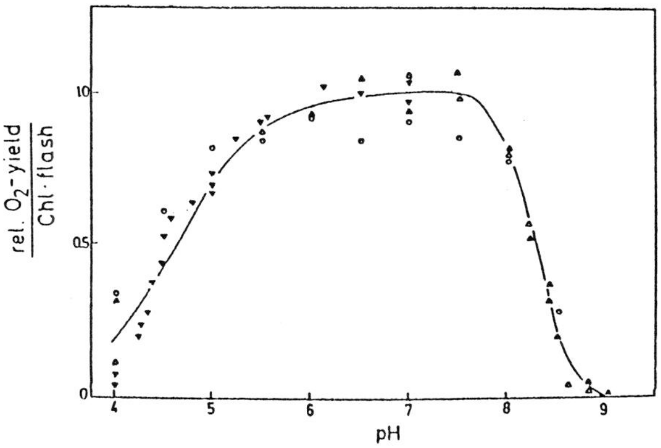

Fig. 7: Relative O_2 yield per flash and chlorophyll in dependence on the pH value. PS II complexes from Synechococcus sp. were excited by saturating Xenon flashes (20 µs FWHM). 10^{-3} M $K_3Fe(CN)_6$ and $2 \cdot 10^{-4}$ M phenyl-p-benzoquinone have been used as artificial electron acceptor.

At high pH O_2 evolution is at a half maximum at about pH 8.3. The reason for the inhibition at high pH might be that Z^+ can no longer induce the $S_1 \rightarrow S_2$ transition. This explanation is supported by an EPR study on the multiline signal that has been assigned to the S_2 state (Dismukes and Siderer, 1981). The multiline EPR-signal amplitude and O_2 evolution show a parallel decrease with increasing pH (Cole et al., 1986).

The kinetics of Z^+ reduction measured by the decay of EPR signal II_{vf} (Babcock et al., 1976) and the concomitant oxidation of the S states (Dekker et al., 1984) coincide and occur with half-life times of about 50 µs ($S_o \rightarrow S_1$), 100 µs ($S_1 \rightarrow S_2$), 100-400 µs ($S_2 \rightarrow S_3$) and 1.3 ms ($S_3 \rightarrow S_o$).

Acknowledgements

I would like to thank the following colleagues for their excellent collaboration during the studies described above: Drs. K. Brettel, J.P. Dekker, S. Gerken, B. Meyer, G.H. Schatz and S. Pauly. I gratefully acknowledge the continuous support provided by Prof. Dr. H.T. Witt.
This work was supported by grants from the Deutsche Forschungsgemeinschaft (Sonderforschungsbereich 312).

Babcock GT (1987) The photosynthetic oxygen evolving process. In: Amesz J (ed) New comprehensive biochemistry, Vol 15, pp 125-158, Elsevier, Amsterdam

Babcock GT, Blankenship RE and Sauer K (1976) Reaction kinetics for positive charge accumulation on the water side of chloroplast photosystem II. FEBS Lett 61:286-289

Barry AB and Babcock GT (1987) Tyrosine radicals are involved in the photosynthetic oxygen-evolving system. Proc Natl Acad Sci USA 84:7099-7103

Bent DV and Hayon E (1975) Excited state chemistry of aromatic amino acids and related peptides. I. Tyrosine. J Am Chem Soc 97:2599-2606

Brettel K, Schlodder E and Witt HT (1984) Nanosecond reduction kinetics of photooxidized chlorpohyll-a_{II} (P-680) in single flashes as a probe for the electron pathway. H^+ release and charge accumulation in the O_2-evolving complex. Biochim Biophys Acta 766:403-428

Bouges-Bocquet B (1980) Kinetic models for the electron donors of photosystem II of photosynthesis. Biochim Biophys Acta 594:85-103

Cole J, Boska M, Blough NV and Sauer K (1986) Reversible and irreversible effects of alkaline pH on Photosystem II electron-transfer reactions. Biochim Biophys Acta 848:41-47

Conjeaud H and Mathis P (1986) Electron transfer in the photosynthetic membrane. Biophys J 49:12-1221

Crofts AR and Wraight CA(1983) The electrochemical domain of photosynthesis. Biochim Biophys Acta 726:149-185

Debus RJ, Barry BA, Babcock GT and McIntosh L (1988) Site-directed mutagenesis identifies a tyrosine radical involved in the photosynthetic oxygen-evolving system. Proc Natl Acad Sci USA 85:427-430

Debus RJ, Barry BA, Sithole I, Babcock GT and McIntosh L (1988) Directed mutagenesis indicates that the donor to $P680^-$ in photosystemII is tyr-161 of the D1 polypeptide. Biochemistry 27:9071-9074

Dekker JP, Plijter JJ, Ouwehand L and van Gorkom HJ (1984) Kinetics of manganese redox transitions in the oxygen-evolving apparatus of photosynthesis. Biochim Biophys Acta 767:176-179

Dekker JP, Boekema EJ, Witt HT and Rögner M (1988) Refined purification and further characterization of oxygen-evolving and Tris-treated photosystem II particles from the thermophilic cyanobacterium Synechococcus sp. Biochim Biophys Acta 936:307-318

Dismukes GC and Siderer Y (1981) Intermediates of a polynuclear manganese center involved in photosynthetic oxidation of water. Proc Natl Acad Sci USA 78:274-278

Döring G, Stiehl HH and Witt HT (1967) A second chlorophyll reaction in the electron chain of photosynthesis. Registration by repetitive excitation technique. Z Naturforsch 22b:639-644

Fowler CF (1977) Proton evolution from photosystem II. Stoichiometry and mechanistic considerations. Biochim Biophys Acta 462:414-421

Förster V and Junge W (1985) Stoichiometry and kinetics of proton release upon photosynthetic water oxidation. Photobiochem Photobiophys 41:183-190

Ghanotakis DF and Yocum CF (1985) Polypeptides of photosystem II and their role in oxygen evolution. Photosynth Res 7:97-114

Gerken S, Brettel K, Schlodder E and Witt HT (1987) Direct observation of the immediate electron donor to chlorophyll-a_{II}^+ (P680$^+$) in oxygen-evolving photosystem II complexes. FEBS Lett 223:376-380

Gerken S, Dekker JP, Schlodder E and Witt HT (1989) Studies on the multiphasic charge recombination between chlorophyll-a_{II}^+ (P680$^+$) and plastoquinone Q_A^- in photosystem II complexes. Ultraviolet difference spectrum of Chla_{II}^+/Chla_{II}. Biochim Biophys Acta

Gerken S, Brettel K, Schlodder E and Witt HT (1988) Optical characterization of the immediate electron donor to chlorophyll a_{II}^+ in O_2-evolving photosystem II complexes. Tyrosine as possible electron carrier between chlorophyll a_{II} and the water-oxidizing manganese complex. FEBS Lett 237:69-75

Hansson Ö and Wydrzynski T (in press) Current perceptions of photosystem II. Photosynth Res

Harriman A (1987) Further comments on the redox potentials of tryptophan and tyrosine. J Phys Chem 91:6102-6104

Homann PH (1987) The relations between the chloride, calcium and polypeptide reequirements of photosynthetic water oxidation. J Bioenerg Biomemr 19:105-123

Joliot P, Barbieri G and Chabaud R (1969) Un nouveau modèle des centres photochimique du système II. Photochem Photobiol 10:309:329

Klimov VV and Krasnovsky AA (1981) Pheophytin as the primary electron acceptor in photosystem 2 reaction centers. Photosynthetics 15:592-609

Kok B, Forbush B and McGloin M (1970) Cooperation of charges in photosynthetic oxygen evolution: I. A linear four step mechanism. Photochem Photobiol 11:457-475

Mathis P and Rutherford AW (1987) The primary reactions of photosystem I and II of algae and higher plants. In: Amesz J (ed) New comprehensive biochemistry, Vol 15, Photosynthesis, pp 63-96. Elsevier, Amsterdam

Meyer B, Schlodder E, Dekker JP and Witt HT (1989) O_2 evolution and Chl-a_{II}^+ (P680$^+$) nanosecond reduction kinetics in single flashes as a function of pH. Biochim Biophys Acta 974:36-43

Meyer B (1987) Thesis, Technische Universität Berlin

Nanba O and Satoh K (1987) Isolation of a photosystem II reaction center consisting of D1 and D2 polypeptides and cytochrome b559. Proc Natl Acad Sci USA 84:109-112

Nuijs AM, van Gorkom HJ, PlijterJJ and Duysens LNM (1986) Primary-charge separation and excitation of chlorophyll a n photosystem II particles from spinach as studied by icosecond absorbance-difference spectroscopy. Biochim iophys Acta 848: 167-175

Pauly S, Schlodder E and Witt HT (in press) No evidence for a specific function of Ca^{2+} and Cl^- inoxygen evolution of isolated PS II complexes from thermophilic cyanobacteria. In: Baltscheffsky M (ed) Proc 8th International Congress on Photosynthesis

Renger G and Völker M (1982) Studies on the proton release pattern of the donor side of system II. FEBS Lett 149:203-207

Satoh K (1985) Protein-pigments and photosystem II reaction center. Photochem Photobiol 42:845-853

Schatz GH and WittHT (1984) Extraction and characterization of oxygen-evolving photosystem II complexes from a thermophilic cyanobacterium Synechococcus sp. Photobiochem Photobiophys 7:1-14

Schatz GH, Broch H and Holzwarth AR (1987) Picosecond kinetics of fluorescence and absorbance changes in photosystem II particles excited at low photon density. Proc Natl Acad Sci USA 84:8414-8418

Schlodder E, Brettel K, Schatz GHand Witt HT (1984) Analysis of the Chl-a_{II}^+ kinetics with nanosecond time resolution in oxygen-evolving photosystem II particles from Synechococcus at 680 and 824 nm. Biochim Biophys Acta 765:178-185

Schlodder E, Brettel K, Witt HT (1985) Relation between microsecond reduction kinetics of photooxidized chlorophyll a_{II} (P-680) and photosynthetic water oxidation. Biochim Biophys Acta 808:123-131

Schlodder E and Meyer B (1987) pH dependence of oxygen evolution and reduction kinetics of photooxidzed chlorophyll-a_{II} (P-680) in photosystem II particles from Synechococcus sp. Biochim Biophys Acta 890:23-31

Schlodder E and Brettel K (1988) Primary charge separation in closed photosystem II with a lifetime of 11 ns. Flash-absorption spectroscopy with O_2-evolving photosystem II complexes from Synechococcus. Biochim Biophys Acta 933:22-34

Stewart AC, Ljungberg U, Akerlund HE and Andersson B (1985) Studies on the polypeptide composition of the cyanobacterial oxygen-evolving xomplex. Biochim Biophys Acta 808:353-362

Stiehl HH and WittHT (1968) Die kurzzeitigen ultravioletten Differenzspektren bei der Photosynthese - Die Redoxreaktion des Plastochinon. Z Naturforsch 23b:220-224

van Best JA and Mathis P (1978) Kinetics of reduction of the oxidized primary electron donor of photosystem II in spinach chloroplasts and in chlorella cells in the microsecond and nanoscond time ranges following flash excitation. Biochim Biophys Acta 503:178-188

Vermaas WFJ, Renger G and Dohnt G (1984) The reduction of the oxygen-evolving system in chloroplasts by thylakoid components. Biochim Biophys Acta 704:194-202

Witt HT, Schlodder E, Brettel K and Saygin Ö (1986) Reaction sequence from light absorption to the cleavage of water in photosynthesis. Photosynthesis Res10:307-326

Yerkes CT, Babcock GT and Crofts AR (1983) A Tris-induced change in the midpoint potential of Z, the donor to photosystem II, as determined by the kinetics of the back reaction. FEBS Lett 158:359-363

Marine biofouling and its control: State-of-the-art and future prospects

L V Evans
Department of Pure and Applied Biology
The University of Leeds
Leeds LS2 9JT
United Kingdom

1. Introduction

Fouling is the result of settlement and growth of micro- or macro-organisms on any man-made or natural submerged substratum. One of the most widely-known effects of fouling in the marine environment is on the efficiency of movement of ships of all kinds, and this paper will be concerned with this aspect. Accumulations of marine growths on the underwater parts of ships will roughen the surface, causing increased friction between hull and water and leading to either an increased use of fuel or loss of speed. The fuel bill for a 250,000 ton supertanker operating 15 knots is approximately £3 million a year; fouling to an extent of only 5% (barely visible) will add some £0.5 million to the fuel costs.

A clean surface in water first spontaneously adsorbs an organic layer of glycoproteins producing a modified substrate which becomes hydrophillic and negatively charged. This is then rapidly colonized by pioneer bacteria whose exudates in turn further modify the surface producing a slime film, to which other bacteria and fouling forms such as invertebrate larvae and algal spores become attached (Characklis & Cooksey, 1983; Baier, 1984). However, because their mode of operation involves short stays in port and rapid transition from temperate to tropical waters, modern supertankers are not generally subject to animal fouling, and algae of different kinds are the principle fouling agents. The commonest of these is the green alga Enteromorpha, which predominates because of its enormous reproductive potential, its capacity to withstand wide fluctuations in environmental conditions such as temperature, salinity and desiccation (Biebl, 1962; Christie & Shaw, 1968), and its rapid and highly effective spore attachment mechanism in response to surface contact (Evans & Christie, 1970; Christie et al., 1970). The brown alga *Ectocarpus* is also a cosmopolitan fouling organism; its spores which are profusely produced germinate and rapidly colonize a surface as a mat of horizontal creeping filaments, from which erect reproductive filaments arise. In addition microalgae, notably diatoms such as *Amphora*, *Achnanthes* and *Amphiprora*, are common fouling forms, being one of the main constituents of slime films (see section 3).

2. Fouling Control

2.1 The Soluble Matrix and Contact Leaching Systems

Conventional methods for controlling fouling involve the inclusion in the paint of toxic compounds which leach out at controlled rates and form a layer of high toxin concentration (the boundary layer) at the paint surface. The thickness of the boundary layer depends on the leaching rate of the toxin, which in turn depends on the temperature, pH and salinity of the seawater and on the flow rate over the surface, being minimal under stationary conditions. For a long time copper compounds, in particular cuprous oxide, have been the main toxic compounds used, incorporated in one of two types of antifouling (AF) composition known as the soluble matrix (free-association) and contact leaching types (see e.g. Evans, 1981). In the former, a binder (vehicle) based on rosin (abietic acid) slowly dissolves in the slightly alkaline seawater, exposing biocide particles, which in turn themselves dissolve in the seawater. Contact leaching AF's contain more copper, in a largely insoluble matrix, and as particles at the surface dissolve deeper ones become exposed. Both AF's are wasteful, since the rate of biocide release is exponential, and there is a high rate of loss in the first 2-3 months after application. Prevention of fouling requires copper to leach at an optimum rate of approximately 10 µg. cm^{-2}. day^{-1}; a rate in excess of this is counterproductive, but a lower leaching rate will result in fouling of the surface. Copper-based systems have a relatively short antifouling life (up to about 15 months) as biocide leaching is retarded by the build up of insoluble salts at the paint surface (Evans, 1981). However, the use of different plasticisers and of blends of rosin with other resins has extended the life of these systems, and also lead to the development of "eroder" compositions, where slight increase in surface roughness occurs during erosion as a result of release of aggregates of vehicle (Hunter, 1989). Notwithstanding this, copper based systems of all kinds are relatively ineffective in the control of *Enteromorpha*.

2.1.1 Settlement of *Enteromorpha* spores

The naked quadriflagellate zoospores, approximately 10 um in length, settle readily in response to surface contact and using EM autoradiography and cytochemistry (Callow & Evans, 1974, 1977) it was shown that attachment occurs by means of a glycoprotein adhesive, present within the anterior end of the spore in electron dense vesicles derived from the Golgi apparatus. The carbohydrate component is synthesised in the Golgi cisternae and this is conjugated with protein from adjacent ER within the Golgi stack. The adhesive is rapidly secreted by settling spores as a fibrillar material (Evans & Christie, 1970) and adhesion improves rapidly during the hours following settlement (Christie *et al*., 1970), presumably due to physical changes taking place in the secreted adhesive. As this happens a cell wall of increasing thickness is also laid down.

2.2 The Diffusion System

Towards the end of the 1960's triorganotins such as triphenyltin chloride (TPTC) were introduced into AF paints, in a system known as the diffusion system (see Evans, 1981). The biocide is molecularly dispersed in a medium with which it is incompatible and diffuses through this, resulting in more diffusing upwards to replace it. Triorganotins are energy transfer inhibitors of both respiration and photosynthesis (see Millner & Evans, 1980, 1981), and if leached into the water at a rate of 2.5-4.0 $\mu g. cm^{-2}. day^{-1}$ they will prevent settlement of many organisms, including Enteromorpha. Ectocarpus and the filamentous green alga Ulothrix however, show considerable resistance to organotin compounds (Mearns, 1973; Millner & Evans, 1980), and Ectocarpus became the predominant fouling alga when organotins were introduced into AF paints in the diffusion system. Release is again exponential, the period of effective AF activity being proportional to the log of the thickness of the AF medium.

2.3 The Copolymer Systems

A major advance in the control of marine algal fouling was made in the early 1970's with the development of the copolymer AF system (see Evans, 1981, 1988). The principal biocide is again a triorganotin, such as tributyltin oxide (TBTO), and hydrolysis of the copolymer results in the controlled release of this. In addition, since surface roughness decreases as this happens, the composition is known as self-polishing copolymer (SPC). Other biocides such as cuprous oxide or cuprous thiocayanate are also incorporated into the polymer matrix and are released during hydrolysis. The AF performance of SPC paints is considerably improved by the incorporation of cuprous (or zinc) oxide (French & Evans, 1986). There is also evidence that when both are used together, the presence of zinc oxide increases the rate of copper release from SPC paints (French _et al._, 1985) and, in addition, laboratory experiments suggest that when used together they are likely to show a synergistic effect against fouling diatoms (French & Evans, 1988). On ships operating under optimal conditions, SPC systems effectively control all macrofouling organisms and these systems are in widespread commercial use, giving an effective antifouling lifetime of up to 4-5 years. However, slime-forming organisms such as diatoms persist, so that it is necessary to improve the effectiveness of SPC AF systems against such organisms.

3. Bacterial/Diatom Slimes

Slimes (or slime films) are composed mainly of bacteria and diatoms and their extracellular products. They can be up to 2mm thick and contain some 30×10^6 bacteria $.g^{-1}$ (Hendey, 1951), and 27×10^4 diatoms $.cm^{-2}$ (Edyvean & Moss, 1988) and the latter have been shown to be highly resistant to copper and/or organotins (see e.g. Daniel & Chamberlain, 1981; Callow & Evans, 1981). In a world-wide survey (Callow, 1986) the most frequently encountered diatoms on AF paints containing only copper, or organotin and copper, were _Amphora_, _Amphiprora_ and _Stauroneis_, whilst _Achnanthes_ species predominated when organotin was the only biocide present, in particular _A. subsessilis_, which is highly resistant to organotins. Slime films, although visually insignificant, have been shown to significantly increase drag. A slime layer 1mm thick on a ship's hull was calculated by Lewthwaite _et al._ (1984) to increase skin friction by 80%, resulting in a 15% loss in speed compared with a slime-free hull.

4. Toxicity Aspects

Since tributyl tin (TBT) was reported to occur in

marinas and harbours at levels in excess of those known to have lethal or sublethal effects on non-target marine organisms (Waldock & Miller, 1983; Cleary & Stebbing, 1985), a ban has been imposed on the use of AF compositions containing TBTO on vessels less than 25 m in length in many countries (Dalley, 1989). Tributyltins are highly toxic because of their ability to penetrate biological membranes. Although they do degrade in seawater to the less toxic dibutyl, monobutyl- and inorganic tin, in turbid waters degradation can take weeks or even months. Thus emphasis has been placed on research aimed in the long term at developing organotin-free paint systems, and in the short term at finding ways of reducing the organotin content of SPC paints, without loss of AF efficiency, or preferably with improved AF efficiency, especially with respect to slime-forming diatoms. Whilst it has not yet been possible to develop an effective tin-free copolymer system which hydrolyses (self-polishes) like the tin-containing SPC system, by the use of biocides which act synergistically in combination with organotin, it may be possible to use reduced levels of the latter, without loss of algicidal activity.

5. Synergistic Effects

5.1 1-dodecylguanidine acetate and 2-dodecylguanidine acetate

The LC50 value (reduction of growth by 50%) of tributyltin oxide (TBTO) to the diatom *Amphora coffeaeformis* is 100 nM, and 1-dodecylguanidine acetate (1-DGA, the fungicide dodine) and 2-dodecylguanidine acetate (2-DGA, a structural isomer of dodine) respectively give LC50 values of 1.25 uM and 13 uM. Both the latter compounds show marked synergism with TBTO, 2-DGA giving the more pronounced response (Fig.1). Thus 10 nM TBTO did not inhibit, and 8 uM 2-DGA by itself gave 25% inhibition, but when used together at these concentrations (Fig.1) 100%

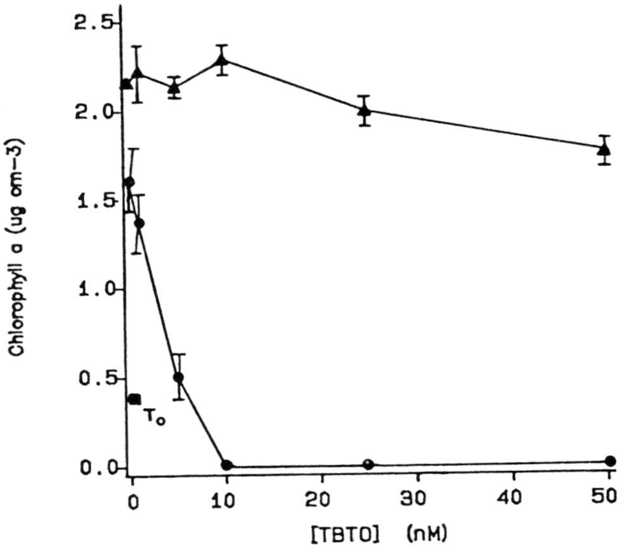

Fig. 1 The effect of tributyltin oxide, 2-DGA, and mixtures of these on the chlorophyll a concentration of *A. coffeaeformis* cultures after 96 h. Tributyltin oxide (▲); tributyltin oxide + 8.0 uM 2-DGA (●). To = chlorophyll concentration when biocide added. Points represent the means of 4 replicates. From Evans *et al*., 1986.)

inhibition occurred (Evans *et al*., 1986). This may be due to increased membrane permeability, caused by the guanidine acetate elevating the intracellular TBTO concentration and resulting in enhanced inhibition of energy transfer even at low exogenous concentrations of TBTO. Maintenance of TBTO and 2-DGA at such synergistically effective concentrations at the boundary layer of an SPC system would in theory permit the organotin level to be markedly reduced, without loss of AF activity. However, this has not yet been achieved in practice.

5. 2 Zineb

The fungicide zineb occurs as a polymeric complex of ethylenebis dithiocarbamate (EBDC) with zinc (Fig. 2). The EBDC ion is believed to degrade in air into further toxic species (Fig. 2). Ethylene thiuram monosulphate (ETM), now revised to 5,6 dihydro-3H-imidazo [2,1-c] -1,2,4-dithiazole-3-thione (DIDT) (Vonk, 1975), has been found to be a major decomposition product (Ludwig et al., 1954; Engst et al., 1968) and ethylene thiuram disulphate (ETD) is an initial intermediate (Hylin, 1973). The toxicity of EBDC's has been attributed to the formation of ethylene diisothiocyanate (EDI) (Kaars-Sijpesteijn et al., 1954; Kaara-Sijpesteijn, 1983; Martin & Woodcock, 1983), which combines unspecifically with cell constituents such as thiol-containing enzymes.

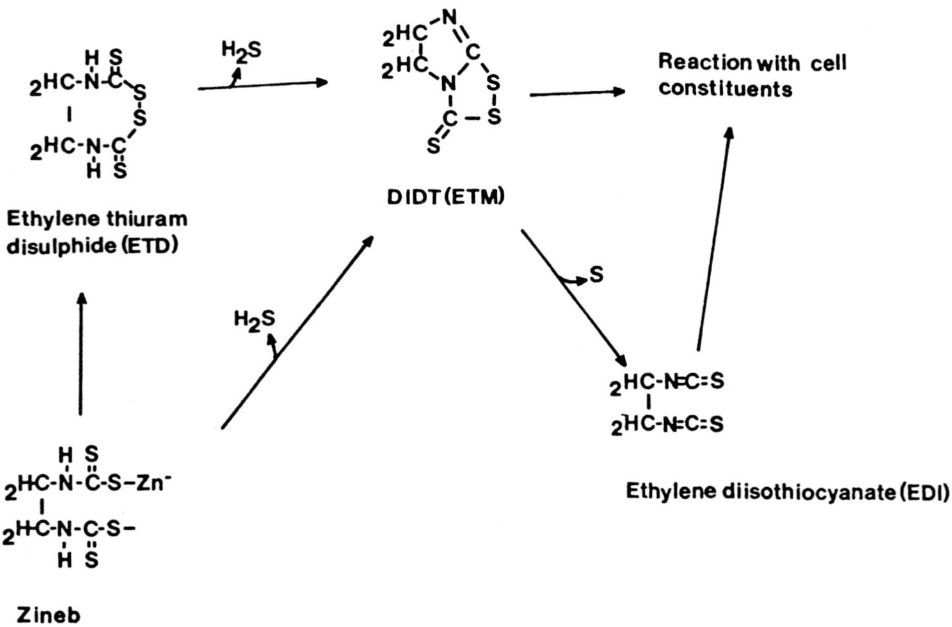

Fig. 2 Some of the decomposition products of zineb (or an EBDC) in water. Adapted by Hunter (1989) from Vonk (1975, 1983).

In a recent study of the toxicity of zineb and its degradation products, singly and in combination with other biocides, using the diatom Amphora coffeaeformis as the test organism (Hunter, 1989; Hunter & Evans, in preparation), zineb and TBTO were found to show a slight antagonistic response. However, a marked synergism was observed between zineb (e.g. at 2 µM) and copper (at concentrations in excess of 33 µM) (Fig. 3) and an even more pronounced synergism (greater than 90%) between DIDT (e.g. at 0.3 µM) and copper (e.g. at 37.8 µM) (Fig. 4). In addition, as can be seen from Table 1, DIDT and EDI were found to be more toxic than zineb, EDI being the most

Table 1. 96h LC_{50} values for zineb and its derivatives with respect to the diatom Amphora coffeaeformis. (From Hunter, 1989.)

Compound	LC_{50} (µM)
Zineb	3.82
DIDT	1.01
EDI	0.38

potent of the compounds tested. However, mixtures of EDI and copper did not show any marked synergism. AF performance was not affected by the use of SPC containing copper and zineb with reduced amounts of TBTO. However, since this did not result in a reduced rate of tin release it is not environmentally beneficial. The most effective combination was zineb and copper in a commercial SPC preparation. The leachate from this was found to contain a compound with an identical retention time to EDI, and copper-mediated degradation of zineb to EDI is postulated to occur, EDI being believed to be responsible for the synergistic effect with copper. Further work on this

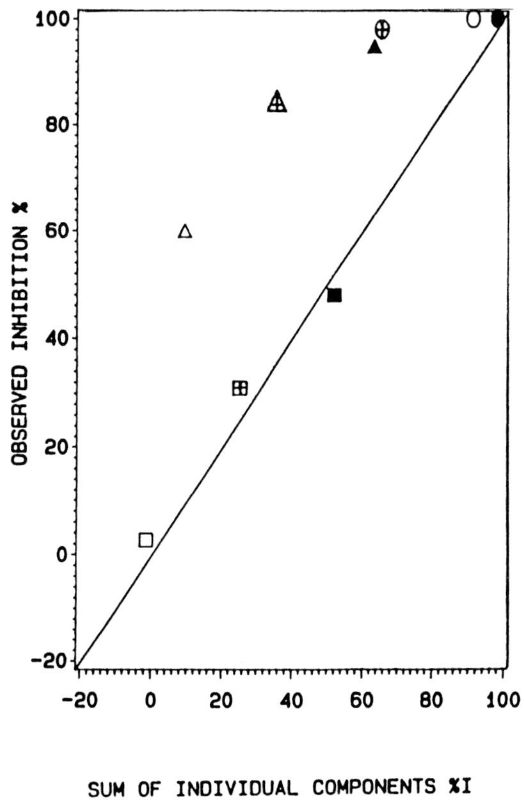

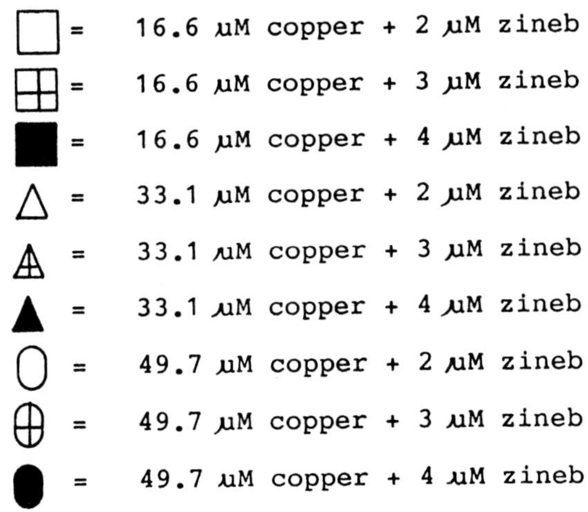

Fig. 3 The effect of copper oxide / zineb mixtures on the chlorophyll a concentration of *A. coffeaeformis* cultures after 96 h, compared to the sum of the individual effects of the biocides tested alone. Points represent the means of 4 replicates. (From Hunter, 1989.)

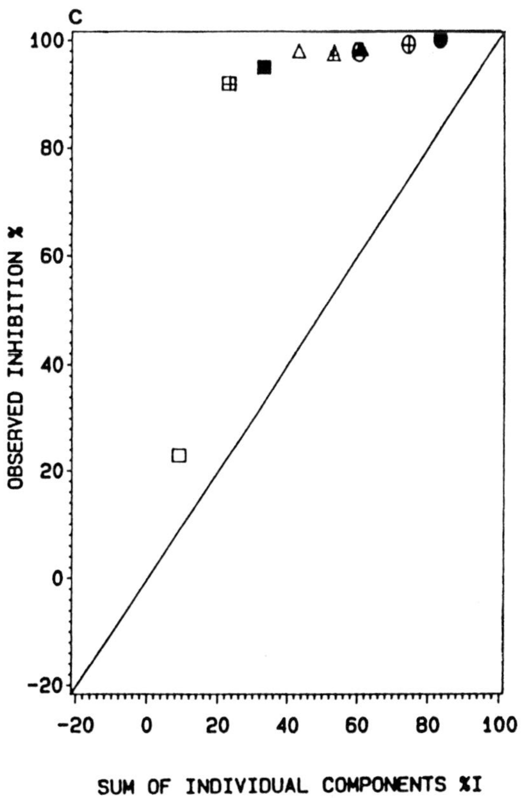

Fig. 4 The effect of copper oxide / DIDT mixtures on the chlorophyll _a_ concentration of _A. coffeaeformis_ cultures after 96 h, compared to the sum of the individual effects of the biocides tested alone. Points represent the means of 4 replicates. (From Hunter, 1989.)

□ = 18.9 µM copper + 0.1 µM DIDT

⊞ = 18.9 µM copper + 0.3 µM DIDT

■ = 18.9 µM copper + 0.6 µM DIDT

△ = 37.8 µM copper + 0.1 µM DIDT

◬ = 37.8 µM copper + 0.3 µM DIDT

▲ = 37.8 µM copper + 0.6 µM DIDT

○ = 59.8 µM copper + 0.1 µM DIDT

⊕ = 59.8 µM copper + 0.3 µM DIDT

● = 59.8 µM copper + 0.6 µM DIDT

synergism is now required, and on the fate of the degradation products.

6. Future prospects

6.1 Biocides

In the longer term, new biocides are needed which have very short half-lives when released, and which degrade into harmless derivatives when they diffuse into the sea away from the AF pain surface. Alternatively, it may be possible to use biocides which are not released from the AF system into the marine environment, and thus cannot affect non-target marine organisms which are not involved in fouling.

6.2 Non-biocidal Control

Another possibility is the use of non-biocidal, non-toxic compounds such as are produced by marine organisms like coral in the prevention of own-surface fouling. Such natural antifoulants have been isolated from sea corals (Rittschoff *et al*., 1985, 1986) and one, a diterpene, inhibits settlement of a barnacle species but not those of a bryozoan, whilst another does the converse. Inhibitory activity is thought to occur by non-toxic modification of the settlement surface, because even at high concentrations the fouling organisms are not killed. However, the apparent species specificity of these compounds (Bakus, 1988) would suggest that their use in a broad spectrum AF system is unlikely.

6.3 Low-energy Surfaces

Although surfaces allowing no adhesion would be difficult to achieve, surfaces to which fouling is only weakly attached woud be easy to clean, using mechanical systems such as soft underwater brushes. Non-toxic, low energy surfaces based for example on silicone elastomers (Milne & Callow, 1985; Callow *et al*., 1986), are being studied, since these markedly weaken adhesion, and they have been applied to test yachts and to deep sea vessels. However, in spite of their good AF performance, their physical properties (e.g. poor abrasion resistance and tear strength) have so far limited their use to coating surfaces such as heat exchangers and buoys.

7. Acknowledgements

The author would like to thank Courtaulds Coatings, International for funding antifouling research at Leeds University over a number of years.

8. References

Bakus GJ (1988) Practical and theoretical problems in the use of fouling panels. In: Thompson M, Sarojini R, Nagabhushanam R (eds), Marine Biodeterioration - Advanced Techniques Applicable to the Indian Ocean. Oxford & IBH Publishing Company, Calcutta, 619-630

Baier RE (1984) Initial events in microbial film formation. In: Costlow JD, Tipper RC (eds), Marine Biodeterioration : An Interdisciplinary Study, U S Naval Research Institute, Annapolis, MD, 57-62

Biebl R (1962) Seaweeds. In: Lewin RA (ed), Physiology and Biochemistry of Algae, Academic Press, 799-815

Callow ME (1986) A world-wide survey of slime formation on antifouling paints. In: Evans LV, Hoagland KD (eds), Algal Biofouling, Elsevier, Amsterdam, 1-20

Callow ME, Evans LV (1974) Studies on the ship-fouling alga Enteromorpha III. Cytochemistry and autoradiography of adhesive production. Protoplasma 80 : 15-27

Callow ME, Evans LV (1977) Studies on the ship-fouling alga Enteromorpha (Chlorophyceae, Ulvales) IV. Polysaccharide and nucleoside diphosphatase localization. Phycologia 16 : 313-320

Callow ME, Evans LV (1981) Some effects of triphenyltin chloride on Achnanthes subsessilis. Bot Mar 24 : 201-205

Callow ME, Pitchers RA, Milne A (1986) The control of fouling by non-biocidal systems. In: Evans LV, Hoagland KD (eds), Algal Biofouling, Elsevier, Amsterdam, 145-158

Characklis WG, Cooksey KE (1983) Biofilms and microbial fouling. Adv appl Microbiol 29 : 93-133

Christie AO, Shaw M (1968) Settlement experiments with zoospores of Enteromorpha intestinalis (L) Link. Br phycol Bull 3 : 529-534

Christie AO, Evans LV, Shaw M (1970) Studies on the ship-fouling alga Enteromorpha intestinalis II. The effect of certain enzymes on the adhesion of zoospores. Ann Bot 34 : 467-482
Christie AO, Evans LV, Shaw M (1970) Studies on the ship-fouling alga Enteromorpha intestinalis II. The effect of certain enzymes on the adhesion of zoospores. Ann Bot 34 : 467-482
Cleary JJ, Stebbing ARD (1985) Organotin and total tin in coastal waters of South West England. Mar Poll Bull 16 : 350-355

Dalley R (1989) Legislation affecting tributyltin antifoulings. Biofouling 1 : 363-366

Daniel GF, Chamberlain AHL (1981) Copper immobilization in fouling diatoms. Bot Mar 24 : 229-243

Edyvean RGJ, Moss BL (1986) Microbial communities on protected steel substrata in sea water. Estuarine Coastal and Shelf Science 22 : 509-527

Engst R, Schnaak W, Rattba H (1968) Fungizide Wirkung und Ruckstandsbildung von Abbauprodukten des Maneb und Zineb an freilandbehandelten Tornaten. NachrBl dt PflSchutzdienst Berl 22 : 26- 29

Evans LV (1981) Marine algae and fouling : a review, with particular reference to ship fouling. Bot Mar 24 : 167-171

Evans LV (1988) Marine biofouling. In: Lembi CA, Waaland JR (eds), Algae and Human Affairs, Cambridge University Press, 433-453

Evans LV, Christie AO (1970) Studies on the ship-fouling alga Enteromorpha intestinalis I. Aspects of the fine structure and biochemistry of swimming and newly settled zoospores. Ann Bot 34 : 451-466

Evans LV, Callow ME, Wood KR (1986) Synergism between antifouling biocides. In: Evans LV, Hoagland KD (eds), Algal Biofouling, Elsevier, Amsterdam, 55-64

French MS, Evans LV (1986) Fouling on paints containing copper and zinc. In: Evans LV, Hoagland KD (eds), Algal Biofouling, Elsevier, Amsterdam, 79-100

French MS, Evans LV (1988) The effects of copper and zinc on growth of the fouling diatoms Amphora and Amphprora. Biofouling 1 : 3-18

French MS, Evans LV, Dalley R (1985) Raft trial experiment on leaching from antifouling paints. In: Polymers in a Marine Environment, Trans I Mar E (C), Vol 97, Conf 2, London, 229-233

Hendey NI (1951) Littoral diatoms of Chichester Harbour with special reference to fouling. Jl R microsc Soc 71 : 1-86

Hunter JE (1989) Synergistic Effects Between Antifouling Biocides. Ph.D. Thesis, University of Leeds

Hylin JW (1973) Oxidative decomposition of ethylene-bis-dithiocarbamate. Bull Env Cont Toxicol 10 : 227-233

Kaars Sijpesteijn A (1983) Mode of action of some traditional fungicides. In: Trinci APJ, Ryley JF (eds), Symp Br Mycol Soc, 135-153

Kaars Sijpesteijn A, Van der Kirk GJM (1954) Investigations on organic fungicides VIII. The biochemical mode of action of bisthiocarbamates and diisothiocyanates. Biochem Biophys Acta 13 : 543-552

Lewthwaite JC, Molland AF, Thomas KW (1984) An investigation into the variation of skin frictional resistance with fouling. Trans R I N A 127 : 269-284

Ludwig RA, Thorn, GD, Unwin CH (1954) Studies on the mechanism of fungicidal action of metallic ethylene bisthiocarbamates. Can J Bot 33 : 42-59

Martin H, Woodcock D (1983) The dithiocarbamate and thiuram derivatives. In: The Scientific Principles of Crop Protection, 7th Edition, Edward Arnold, 130-135

Mearns RD (1973) Vinyl weed-resistant antifouling : a practical approach to laboratory formulation. J Oil Col Chem Ass 50 : 201-208

Millner PA, Evans LV (1980) The effects of triphenyltin chloride on respiration and photosynthesis in the green algae *Enteromorpha intestinalis* and *Ulothrix pseudoflacca*. Plant, Cell and Environment 3 : 339-348

Millner PA, Evans LV (1981) Uptake of triphenyltin chloride by *Enteromorpha intestinalis* and *Ulothrix pseudoflacca*. Plant, Cell and Environment 4 : 383-389

Milne A, Callow ME (1985) Non-biocidal antifouling processes. In: Polymers in a Marine Environment. Trans I Mar E (C), Vol 97, Conf 2, London, 229-233

Rittschoff D, Hooper IR, Branscomb ES, Costlow JD (1985) Inhibition of barnacle settlement and behaviour by natural products from whip corals, *L,. virgulata* (Lamarck, 1815). J Chem Ecol 11 : 551-563

Rittschoff D, Hooper IR, Costlow JD (1986) Barnacle settlement inhibitors from sea pansies (*Renilla reniformis*). Bull Mar Sci 39 : 376-382

Vonk JW (1975) Chemical Composition of Bisdithiocarbamate Fungicides and their Metabolism by Plants and Animals. Ph.D Thesis TNO, Utrecht

Vonk JW (1983) Metabolism of fungicides in plants. In: Hutson DH, Roberts TR (eds), Pesticide Biochemistry and Toxicology. John Wiley & Sons, 111-157

Waldock MJ, Miller D (1983) The determination of total tributyltin in seawater and oysters in areas of high pleasure craft activity. ICES paper CM E : 12, International Council for the Exploration of the Sea, Copenhagen

Taxonomic Index

Subject Index